Hands-On Computer Vision with Detectron2

Develop object detection and segmentation models with a code and visualization approach

Van Vung Pham

BIRMINGHAM—MUMBAI

Hands-On Computer Vision with Detectron2

Copyright © 2023 Packt Publishing

Publishing Product Manager: Dhruv J. Kataria
Content Development Editor: Shreya Moharir
Technical Editor: Rahul Limbachiya
Copy Editor: Safis Editing
Project Coordinator: Farheen Fathima
Proofreader: Safis Editing
Indexer: Pratik Shirodkar
Production Designer: Jyoti Chauhan
Marketing Coordinators: Shifa Ansari, Vinishka Kalra

First published: April 2023

Production reference: 1290323

Published by Packt Publishing Ltd.
Livery Place
35 Livery Street
Birmingham
B3 2PB, UK.

ISBN 978-1-80056-162-5

www.packtpub.com

Foreword

I have known and worked with Van Vung Pham for more than 10 years and was also his academic advisor for his doctoral degree. Vung won several data visualization, computer vision, and machine learning challenges during his Ph.D. program, including using Detectron2 to detect and classify road damage. In this book, *Hands-On Computer Vision with Detectron2*, Vung takes you on a learning journey that starts with common computer vision tasks. He then walks you through the steps for developing computer vision applications using stunning deep-learning models with simple code by utilizing pre-trained models on the Detectron2 Model Zoo.

Existing models, trained on huge datasets, and for the most common object types, can meet common computer vision tasks. However, this book also focuses on developing computer vision applications on a custom domain for specific business requirements. For this, Vung provides the steps to collect and prepare data, train models, and fine-tune models on brain tumor datasets for object detection and instance segmentation tasks to illustrate how to develop computer vision applications on custom business domains.

In his presentations and examples, Vung provides code that can be conveniently executed on Google Colab and visualizations to help illustrate theoretical concepts. The ability to execute the code on Google Colab helps eliminate the burden of hardware and software setup, so you can get started quickly and conveniently. The visualizations allow you to easily grasp complicated computer vision concepts, better understand deep learning architectures for computer vision tasks, and become an expert in this area.

Beyond developing deep learning models for computer vision tasks, you will learn how to deploy the trained models to various environments. Vung explains different model formats, such as TorchScript and ONNX formats, and their respective execution platforms and environments, such as C++ servers, web browsers, or mobile and edge devices.

Become a developer and an expert in developing and deploying computer vision applications with Detectron2.

– Tommy Dang

iDVL director and assistant professor, Texas Tech University

Contributors

About the author

Van Vung Pham is a passionate research scientist in machine learning, deep learning, data science, and data visualization. He has years of experience and numerous publications in these areas. He is currently working on projects that use deep learning to predict road damage from pictures or videos taken from roads. One of the projects uses Detectron2 and Faster R-CNN to predict and classify road damage and achieve state-of-the-art results for this task. Dr. Pham obtained his Ph.D. from the Computer Science Department, at Texas Tech University, Lubbock, Texas, USA. He is currently an assistant professor at the Computer Science Department, Sam Houston State University, Huntsville, Texas, USA.

I want to thank the people who have been close and supported me, especially my wife, Hanh, my parents, my children, and my Ph.D. advisor (Dr. Tommy Dang from Texas Tech University).

About the reviewers

Yiqiao Yin is a senior data scientist at an S&P 500 company LabCorp, developing AI-driven solutions for drug diagnostics and development. He has a BA in mathematics and a BSc in finance from the University of Rochester. He was a PhD student in statistics at Columbia University and has a wide range of research interests in representation learning: feature learning, deep learning, computer vision, and natural language processing. He has held professional positions as an enterprise-level data scientist at EURO STOXX 50 company Bayer, a quantitative researcher at AQR, working on alternative quantitative strategies to portfolio management and factor-based trading, and an equity trader at T3 Trading on Wall Street.

Nikita Dalvi is a highly skilled and experienced technical professional, currently pursuing a master's degree in computing and data science at Sam Houston State University. With a background in information and technology, she has honed her skills in programming languages such as Java and Python over the past five years, having worked with prestigious organizations such as Deloitte and Tech Mahindra. Driven by her passion for programming, she has taught herself new languages and technologies over the years and stayed up to date with the latest industry trends and best practices.

Table of Contents

Part 2: Developing Custom Object Detection Models

3

Data Preparation for Object Detection Applications   43

Part 3: Developing a Custom Detectron2 Model for Instance Segmentation Tasks

10

11

Part 4: Deploying Detectron2 Models into Production

12

13

Deploying Detectron2 Models into Browsers and Mobile Environments   265

Preface

Computer vision takes part and has become a critical success factor in many modern businesses such as automobile, robotics, manufacturing, and biomedical image processing – and its market is growing rapidly. This book will help you explore **Detectron2**. It is the next-generation library that provides cutting-edge computer vision algorithms. Many research and practical projects at Facebook (now Meta) use it as a library to support computer vision tasks. Its models can be exported to TorchScript and Open Neural Network Exchange (ONNX) format for deployments into server production environments (such as C++ runtime), browsers, and mobile devices.

By utilizing code and visualizations, this book will guide you on using existing models in Detectron2 for computer vision tasks (object detection, instance segmentation, key-point detection, semantic detection, and panoptic segmentation). It also covers theories and visualizations of Detectron2's architectures and how each module in Detectron2 works. This book walks you through two complete hands-on, real-life projects (preparing data, training models, fine-tuning models, and deployments) for object detection and instance segmentation of brain tumors using Detectron2.

The data preparation section discusses common sources of datasets for computer vision applications and tools to collect and label data. It also describes common image data annotation formats and codes to convert from different formats to the one Detectron2 supports. The training model section guides the steps to prepare the configuration file, load pre-trained weights for transfer learning (if necessary), and modify the default trainer to meet custom business requirements.

The fine-tuning model section includes inspecting training results using TensorBoard and optimizing Detectron2 solvers. It also provides a primer to common and cutting-edge image augmentation techniques and how to use existing Detectron2 image augmentation techniques or to build and apply custom image augmentation techniques at training and testing time. There are also techniques to fine-tune object detection models, such as computing appropriate configurations for generating anchors (sizes and ratios of the anchors) or means or standard deviations of the pixel values from custom datasets. For instance segmentation task, this book also discusses the use of PointRend to improve the quality of the boundaries of the detected instances.

This book also covers steps for deploying Detectron2 models into production and developing Detectron2 applications for mobile devices. Specifically, it provides the model formats and platforms that Detectron2 supports, such as TorchScript and ONNX formats. It provides the code to convert Detectron2 into these formats models using tracing and scripting approaches. Additionally, code snippets illustrate how to deploy Detectron2 models into C++ and browser environments. Finally, this book also discusses D2Go, a platform to train, fine-tune, and quantize computer visions so they can be deployable to mobile and edge devices with low-computation resource awareness.

Through this book, you will find that Detectron2 is a valuable framework for anyone looking to build robust computer vision applications.

Who this book is for

If you are a deep learning application developer, researcher, or software developer with some prior knowledge about deep learning, this book is for you to get started and develop deep learning models for computer vision applications. Even if you are an expert in computer vision and curious about the features of Detectron2, or you would like to learn some cutting-edge deep learning design patterns, you will find this book helpful. Some HTML, Android, and C++ programming skills are advantageous if you want to deploy computer vision applications using these platforms.

What this book covers

Chapter 1, An Introduction to Detectron2 and Computer Vision Tasks, introduces Detectron2, its architectures, and the computer vision tasks that Detectron2 can perform. Additionally, this chapter provides the steps to set up environments for developing computer vision applications using Detectron2.

Chapter 2, Developing Computer Vision Applications Using Existing Detectron2 Models, guides you through the steps to develop applications for computer vision tasks using state-of-the-art models in the Detectron2 Model Zoo. Thus, you can quickly develop practical computer vision applications without having to train custom models.

Chapter 3, Data Preparation for Object Detection Applications, discusses the steps to prepare data for training models using Detectron2. Additionally, this chapter covers the techniques to convert standard annotation formats to the data format required by Detectron2 in case the existing datasets come in different formats.

Chapter 4, The Architecture of the Object Detection Model in Detectron2, dives deep into the architecture of Detectron2 for the object detection task. This chapter is essential for understanding common terminologies when designing deep neural networks for vision systems.

Chapter 5, Training Custom Object Detection Models, provides steps to prepare data, train an object detection model, select the best model, and perform inferencing object detection tasks. Additionally, it details the development process of a custom trainer by extending the default trainer and incorporating a hook into the training process.

Chapter 6, Inspecting Training Results and Fine-Tuning Detectron2's Solver, covers the steps to use TensorBoard to inspect training histories. It utilizes the codes and visualizations approach for explaining the concepts behind Detectron2's solvers and their hyperparameters. The related concepts include gradient descent, Stochastic gradient descent, momentum, and variable learning rate optimizers.

Chapter 7, Fine-Tuning Object Detection Models, explains how Detectron2 processes its inputs and provides codes to analyze the ground-truth boxes from a training dataset and find appropriate values for the anchor sizes and ratio configuration parameters. Additionally, this chapter provides the code to calculate the input image pixels' means and standard deviations from the training dataset in a rolling manner. The rolling calculations of these hyperparameters are essential if the training dataset is large and does not fit in the memory.

Chapter 8, Image Data Augmentation Techniques, introduces Detectron2's image augmentation system with three main components: Transformation, Augmentation, and AugInput. It describes classes in these components and how they work together to perform image augmentation while training Detectron2 models.

Chapter 9, Applying Train-Time and Test-Time Image Augmentations, introduces the steps to apply these existing classes to training. This chapter also explains how to modify existing codes to implement custom techniques that need to load data from different inputs. Additionally, this chapter details the steps for applying image augmentations during test time to improve accuracy.

Chapter 10, Training Instance Segmentation Models, covers the steps to construct a dataset in the format supported by Detectron2 and train a model for a segmentation task. This chapter also utilizes the codes and visualizations approach to explain the architecture of an object segmentation application developed using Detectron2.

Chapter 11, Fine-Tuning Instance Segmentation Models, introduces PointRend, a project inside Detectron2 that helps improve the sharpness of the object's boundaries. This chapter also covers the steps to use existing PointRend models and to train custom models using PointRend.

Chapter 12, Deploying Detectron2 Models into Server Environments, walks you through the steps in an export process to convert Detectron2 models into deployable artifacts. This chapter then provides the steps to deploy the exported models into the server environments.

Chapter 13, Deploying Detectron2 Models into Browsers and Mobile Environments, introduces the ONNX framework. It is extremely helpful when deploying Detectron2 models into browsers or mobile environments is needed. This chapter also describes D2Go for training, quantizing lightweight models extremely useful for deploying into mobile or edge devices.

To get the most out of this book

Detectron2, D2Go, and PyTorch are under active development, and therefore Detectron2 or D2Go may not be compatible with the PyTorch version you have or that Google Colab provides by default. The source code is fully tested using the following versions on Google Colab:

Software/hardware covered in the book	Operating system requirements
Python 3.8 and 3.9	Google Colab
PyTorch 1.13	
CUDA: cu116	
Detectron2 (commit 3ed6698)	
D2Go (commit 1506551)	

Chapter 1 of this book also provides installation instructions and information you need to start. Additionally, this book provides Code in Action videos where you can view the Python and commit versions of all the packages being used.

If you are using the digital version of this book, we advise you to type the code yourself or access the code from the book's GitHub repository (a link is available in the next section). Doing so will help you avoid any potential errors related to the copying and pasting of code.

Note that some less important portions of the codes are truncated inside the book for space efficiency and legibility. Therefore, simply copying and pasting codes from the book may lead to execution errors. It is recommended to follow the complete code found in the book's GitHub repository, detailed in the following section.

Download the example code files

You can download the example code files for this book from GitHub at `https://github.com/PacktPublishing/Hands-On-Computer-Vision-with-Detectron2`. If there's an update to the code, it will be updated in the GitHub repository.

We also have other code bundles from our rich catalog of books and videos available at `https://github.com/PacktPublishing/`. Check them out!

Code in Action

The Code in Action videos for this book can be viewed at `http://bit.ly/40DJdpd`.

Conventions used

There are a number of text conventions used throughout this book.

`Code in text`: Indicates code words in text, database table names, folder names, filenames, file extensions, pathnames, dummy URLs, user input, and Twitter handles. Here is an example: "Mount the downloaded `WebStorm-10*.dmg` disk image file as another disk in your system."

A block of code is set as follows:

```
html, body, #map {
  height: 100%;
  margin: 0;
  padding: 0
}
```

When we wish to draw your attention to a particular part of a code block, the relevant lines or items are set in bold:

```
[default]
exten => s,1,Dial(Zap/1|30)
exten => s,2,Voicemail(u100)
exten => s,102,Voicemail(b100)
exten => i,1,Voicemail(s0)
```

Any command-line input or output is written as follows:

```
$ mkdir css
$ cd css
```

Bold: Indicates a new term, an important word, or words that you see onscreen. For instance, words in menus or dialog boxes appear in **bold**. Here is an example: "Select **System info** from the **Administration** panel."

> Tips or important notes
> Appear like this.

Get in touch

Feedback from our readers is always welcome.

General feedback: If you have questions about any aspect of this book, email us at `customercare@packtpub.com` and mention the book title in the subject of your message.

Errata: Although we have taken every care to ensure the accuracy of our content, mistakes do happen. If you have found a mistake in this book, we would be grateful if you would report this to us. Please visit `www.packtpub.com/support/errata` and fill in the form.

Piracy: If you come across any illegal copies of our works in any form on the internet, we would be grateful if you would provide us with the location address or website name. Please contact us at `copyright@packt.com` with a link to the material.

If you are interested in becoming an author: If there is a topic that you have expertise in and you are interested in either writing or contributing to a book, please visit `authors.packtpub.com`.

Share Your Thoughts

Once you've read *Hands-On Computer Vision with Detectron2*, we'd love to hear your thoughts! Scan the QR code below to go straight to the Amazon review page for this book and share your feedback.

https://packt.link/r/1-800-56162-8

Your review is important to us and the tech community and will help us make sure we're delivering excellent quality content.

Download a free PDF copy of this book

Thanks for purchasing this book!

Do you like to read on the go but are unable to carry your print books everywhere?

Is your eBook purchase not compatible with the device of your choice?

Don't worry, now with every Packt book you get a DRM-free PDF version of that book at no cost.

Read anywhere, any place, on any device. Search, copy, and paste code from your favorite technical books directly into your application.

The perks don't stop there, you can get exclusive access to discounts, newsletters, and great free content in your inbox daily

Follow these simple steps to get the benefits:

1. Scan the QR code or visit the link below

https://packt.link/free-ebook/9781800561625

2. Submit your proof of purchase
3. That's it! We'll send your free PDF and other benefits to your email directly

Part 1: Introduction to Detectron2

This first part introduces Detectron2, its architectures, and the computer vision tasks that Detectron2 can perform. In other words, it discusses why we need computer vision applications and what computer vision tasks Detectron2 can perform. Additionally, this part provides the steps to set up environments for developing computer vision applications using Detectron2 locally or on the cloud using Google Colab. Also, it guides you through the steps to build applications for computer vision tasks using state-of-the-art models in Detectron2. Specifically, it discusses the existing and pre-trained models in Detectron2's Model Zoo and the steps to develop applications for object detection, instance segmentation, key-point detection, semantic segmentation, and panoptic segmentation using these models.

The first part covers the following chapters:

- *Chapter 1, An Introduction to Detectron2 and Computer Vision Tasks*
- *Chapter 2, Developing Computer Vision Applications Using Existing Detectron2 Models*

1

An Introduction to Detectron2 and Computer Vision Tasks

This chapter introduces **Detectron2**, its architectures, and the **computer vision** (**CV**) tasks that Detectron2 can perform. In other words, this chapter discusses what CV tasks Detectron2 can perform and why we need them. Additionally, this chapter provides the steps to set up environments for developing CV applications using Detectron2 locally or on the cloud using Google Colab.

By the end of this chapter, you will understand the main CV tasks (e.g, object detection, instance segmentation, keypoint detection, semantic segmentation, and panoptic segmentation); know how Detectron2 works and what it can do to help you tackle CV tasks using **deep learning**; and be able to set up local and cloud environments for developing Detectron2 applications.

Specifically, this chapter covers the following topics:

- Computer vision tasks
- Introduction to Detectron2 and its architecture
- Detectron2 development environments

Technical requirements

Detectron2 CV applications are built on top of **PyTorch**. Therefore, a compatible version of PyTorch is expected to run the code examples in this chapter. Later sections of this chapter will provide setup instructions specifically for Detectron2. All the code, datasets, and respective results are available on the GitHub page of the book at `https://github.com/PacktPublishing/Hands-On-Computer-Vision-with-Detectron2`. It is highly recommended to download the code and follow along.

Computer vision tasks

Deep learning achieves state-of-the-art results in many CV tasks. The most common CV task is image classification, in which a deep learning model gives a class label for a given image. However, recent advancements in deep learning allow computers to perform more advanced vision tasks. There are many of these advanced vision tasks.

However, this book focuses on more common and important ones, including object detection, instance segmentation, keypoint detection, semantic segmentation, and panoptic segmentation. It might be challenging for readers to differentiate between these tasks. *Figure 1.1* depicts the differences between them. This section outlines *what* they are and *when* to use them, and the rest of the book focuses on *how* to implement these tasks using Detectron2. Let's get started!

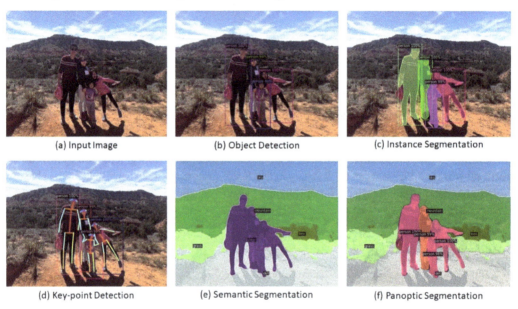

Figure 1.1: Common computer vision tasks

Object detection

Object detection generally includes object localization and classification. Specifically, deep learning models for this task predict where objects of interest are in an image by applying the bounding boxes around these objects (*localization*). Furthermore, these models also classify the detected objects into types of interest (*classification*).

One example of this task is specifying people in pictures and applying bounding boxes to the detected humans (*localization only*), as shown in *Figure 1.1 (b)*. Another example is to detect road damage from a recorded road image by providing bounding boxes to the damage (localization) and further classifying the damage into types such as longitudinal cracks, traverse cracks, alligator cracks, and potholes (classification).

Instance segmentation

Like object detection, instance segmentation also involves object localization and classification. However, instance segmentation takes things one step further while localizing the detected objects of interest.

Specifically, besides classification, models for this task localize the detected objects at the *pixel level*. In other words, it identifies all the pixels of each detected object. Instance segmentation is needed in applications that require shapes of the detected objects in images and need to track every individual object. *Figure 1.1 (c)* shows the instance segmentation result on the input image in *Figure 1.1 (a)*. Specifically, besides the bounding boxes, every pixel of each person is also highlighted.

Keypoint detection

Besides detecting objects, keypoint detection also indicates important parts of the detected objects called keypoints. These keypoints describe the detected object's essential trait. This trait is often invariant to image rotation, shrinkage, translation, or distortion. For instance, the keypoints of humans include the eyes, nose, shoulders, elbows, hands, knees, and feet. Keypoint detection is important for applications such as action estimation, pose detection, or face detection. *Figure 1.1 (d)* shows the keypoint detection result on the input image in *Figure 1.1 (a)*. Specifically, besides the bounding boxes, it highlights all keypoints for every detected individual.

Semantic segmentation

A semantic segmentation task does not detect specific instances of objects but classifies each pixel in an image into some classes of interest. For instance, a model for this task classifies regions of images into pedestrians, roads, cars, trees, buildings, and the sky in a self-driving car application. This task is important when providing a broader view of groups of objects with different classes (i.e., a higher level of understanding of the image). Specifically, if individual class instances are in one region, they are grouped into one mask instead of having a different mask for each individual.

One example of the application of semantic segmentation is to segment the images into foreground objects and background objects (e.g., to blur the background and provide a more artistic look for a portrait image). *Figure 1.1 (e)* shows the semantic segmentation result on the input image in *Figure 1.1 (a)*. Specifically, the input picture is divided into regions classified as things (people or front objects) and background objects such as the sky, a mountain, dirt, grass, and a tree.

Panoptic segmentation

Panoptic literally means *"everything visible in the image"*. In other words, it can be viewed as combining common CV tasks such as instance segmentation and semantic segmentation. It helps to show the unified and global view of segmentation. Generally, it classifies objects in an image into foreground objects (that have proper geometries) and background objects (that do not have appropriate geometries but are textures or materials).

Examples of foreground objects include people, animals, and cars. Likewise, examples of background objects include the sky, dirt, trees, mountains, and grass. Different from semantic segmentation, panoptic segmentation does not group consecutive individual objects of the same class into one region. *Figure 1.1 (f)* shows the panoptic segmentation result on the input image in *Figure 1.1 (a)*.

Specifically, it looks similar to the semantic segmentation result, except it highlights the individual instances separately.

> **Important note – other CV tasks**
>
> There are other advanced CV projects developed on top of Detectron2, such as DensePose and PointRend. However, this book focuses on developing CV applications for the more common ones, including object detection, instance segmentation, keypoint detection, semantic segmentation, and panoptic segmentation in *Chapter 2*. Furthermore, *Part 2* and *Part 3* of this book further explore developing custom CV applications for the two most important tasks (object detection and instance segmentation). There is also a section that describes how to use PointRend to improve instance segmentation quality. Additionally, it is relatively easy to expand the code for other tasks once you understand these tasks.

Let's get started by getting to know Detectron2 and its architecture!

An introduction to Detectron2 and its architecture

Detectron2 is Facebook (now Meta) AI Research's open source project. It is a next-generation library that provides cutting-edge detection and segmentation algorithms. Many research and practical projects at Facebook use it as a library to support implementing CV tasks. The following sections introduce Detectron2 and provide an overview of its architecture.

Introducing Detectron2

Detectron2 implements state-of-the-art detection algorithms, such as Mask R-CNN, RetinaNet, Faster R-CNN, RPN, TensorMask, PointRend, DensePose, and more. The question that immediately comes to mind after this statement is, why is it better if it re-implements existing cutting-edge algorithms? The answer is that Detectron2 has the advantages of being faster, more accurate, modular, customizable, and built on top of PyTorch.

Specifically, it is *faster* and more accurate because while reimplementing the cutting-edge algorithms, there is the chance that Detectron2 will find suboptimal implementation parts or obsolete features from older versions of these algorithms and re-implement them. It is *modular*, or it divides its implementation into sub-parts. The parts include the input data, backbone network, region proposal heads, and prediction heads (the next section covers more information about these components). It is *customizable*, meaning its components have built-in implementations, but they can be customized by calling new implementations. Finally, it is built on top of PyTorch, meaning that many developer resources are available online to help develop applications with Detectron2.

Furthermore, Detectron2 provides *pre-trained models* with state-of-the-art detection results for CV tasks. These models were trained with many images on high computation resources at the Facebook research lab that might not be available in other institutions.

These pre-trained models are published on its Model Zoo and are free to use: `https://github.com/facebookresearch/detectron2/blob/main/MODEL_ZOO.md`.

These pre-trained models help developers develop typical CV applications quickly without collecting, preparing many images, or requiring high computation resources to train new models. However, suppose there is a need for developing a CV task on a specific domain with a custom dataset. In that case, these existing models can be the starting weights, and the whole Detectron2 model can be trained again on the custom dataset.

Finally, we can convert Detectron2 models into *deployable artifacts*. Precisely, we can convert Detectron2 models into standard file formats of standard deep learning frameworks such as `TorchScript`, `Caffe2 protobuf`, and `ONNX`. These files can then be deployed to their corresponding runtimes, such as `PyTorch`, `Caffe2`, and `ONNX Runtime`. Furthermore, Facebook AI Research also published **Detectron2Go (D2Go)**, a platform where developers can take their Detectron2 development one step further and create models optimized for mobile devices.

In summary, Detectron2 implements cutting-edge detection algorithms with the advantage of being fast, accurate, modular, and built on top of PyTorch. Detectron2 also provides pre-trained models so users can get started and quickly build CV applications with state-of-the-art results. It is also customizable, so users can change its components or train CV applications on a custom business domain. Furthermore, we can export Detectron2 into scripts supported by standard deep learning framework runtimes. Additionally, initial research called Detectron2Go supports developing Detectron2 applications for edge devices.

In the next section, we will look into Detectron2 architecture to understand how it works and the possibilities of customizing each of its components.

Detectron2 architecture

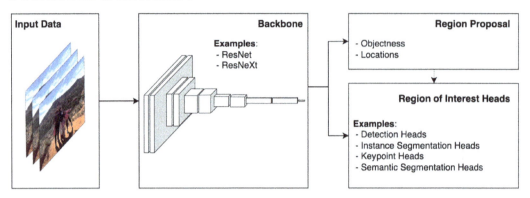

Figure 1.2: The main components of Detectron2

Detectron2 has a modular architecture. *Figure 1.2* depicts the four main modules in a standard Detectron2 application. The first module is for registering input data (**Input Data**).

The second module is the backbone to extract image features (**Backbone**), followed by the third one for proposing regions with and without objects to be fed to the next training stage (**Region Proposal**). Finally, the last module uses appropriate heads (such as detection heads, instance segmentation heads, keypoint heads, semantic segmentation heads, or panoptic heads) to predict the regions with objects and classify detected objects into classes. *Chapter 3* to *Chapter 5* discuss these components for building a CV application for object detection tasks, and *Chapter 10 and Chapter 11* detail these components for segmentation tasks. The following sections briefly discuss these components in general.

The input data module

The input data module is designed to load data in large batches from hard drives with optimization techniques such as caching and multi-workers. Furthermore, it is relatively easy to plug data augmentation techniques into a data loader for this module. Additionally, it is designed to be customizable so that users can register their custom datasets. The following is the typical syntax for assigning a custom dataset to train a Detectron2 model using this module:

```
DatasetRegistry.register(
    'my_dataset',
    load_my_dataset
)
```

The backbone module

The backbone module extracts features from the input images. Therefore, this module often uses a cutting-edge convolutional neural network such as ResNet or ResNeXt. This module can be customized to call any standard convolutional neural network that performs well in an image classification task of interest. Notably, this module has a great deal of knowledge about transfer learning. Specifically, we can use those pre-trained models here if we want to use a state-of-the-art convolution neural network that works well with large image datasets such as ImageNet. Otherwise, we can choose those simple networks for this module to increase performance (training and prediction time) with the accuracy trade-off. *Chapter 2* will discuss selecting appropriate pre-trained models on the Detectron2 Model Zoo for common CV tasks.

The following code snippet shows the typical syntax for registering a custom backbone network to train the Detectron2 model using this module:

```
@BACKBONE_REGISTRY.register()
class CustomBackbone(Backbone):
    pass
```

The region proposal module

The next module is the region proposal module (**Region Proposal**). This module accepts the extracted features from the backbone and predicts or proposes image regions (with location specifications) and scores to indicate whether the regions contain objects (with objectness scores). The objectness score of a proposed region may be 0 (for not having an object or being background) or 1 (for being sure that there is an object of interest in the predicted region). Notably, this object score is not about the probability of being a class of interest but simply whether the region contains an object (of any class) or not (background).

This module is set with a default **Region Proposal Network** (**RPN**). However, replacing this network with a custom one is relatively easy. The following is the typical syntax for registering a custom RPN to train the Detectron2 model using this module:

```
@ROI_BOX_HEAD_REGISTRY.register()
class CustomBoxHead(nn.Module):
    pass
```

Region of interest module

The last module is the place for the **region of interest** (**RoI**) heads. Depending on the CV tasks, we can select appropriate heads for this module, such as detection heads, segmentation heads, keypoint heads, or semantic segmentation heads. For instance, the detection heads accept the region proposals and the input features of the proposed regions and pass them through a fully connected network, with two separate heads for prediction and classification. Specifically, one head is used to predict bounding boxes for objects, and another is for classifying the detected bounding boxes into corresponding classes.

On the other hand, semantic segmentation heads also use convolutional neural network heads to classify each pixel into one of the classes of interest. The following is the typical syntax for registering custom region of interest heads to train the Detectron2 model using this module:

```
@ROI_HEAD_REGISTRY.register()
class CustomHeads(StandardROIHeads):
    pass
```

Now that you have an understanding of Detectron2 and its architecture, let's prepare development environments for developing Detectron2 applications.

Detectron2 development environments

Now, we understand the advanced CV tasks and how Detectron2 helps to develop applications for these tasks. It is time to start developing Detectron2 applications. This section provides steps to set up Detectron2 development environments on the cloud using Google Colab, a local environment, or a hybrid approach connecting Google Colab to a locally hosted runtime.

Cloud development environment for Detectron2 applications

Google Colab or **Colaboratory** (https://colab.research.google.com) is a cloud platform that allows you to write and execute Python code from your web browser. It enables users to start developing deep learning applications with zero configuration because most common machine learning and deep learning packages, such as PyTorch and TensorFlow, are pre-installed. Furthermore, users will have access to GPUs free of charge. Even with the free plan, users have access to a computation resource that is relatively better than a standard personal computer. Users can pay a small amount for Pro or Pro+ with higher computation resources if needed. Additionally, as its name indicates, it is relatively easy to collaborate on Google Colab, and it is easy to share Google Colab files and projects.

Deep learning models for CV tasks work with many images; thus, GPUs significantly speed up the training and inferencing time. However, by default, Google Colab does not enable GPUs' runtime. Therefore, users should enable the GPU hardware accelerator before installing Detectron2 or training Detectron2 applications. This step is to select **GPU** from the **Hardware accelerator** drop-down menu found under **Runtime | Change runtime type**, as shown in *Figure 1.3*:

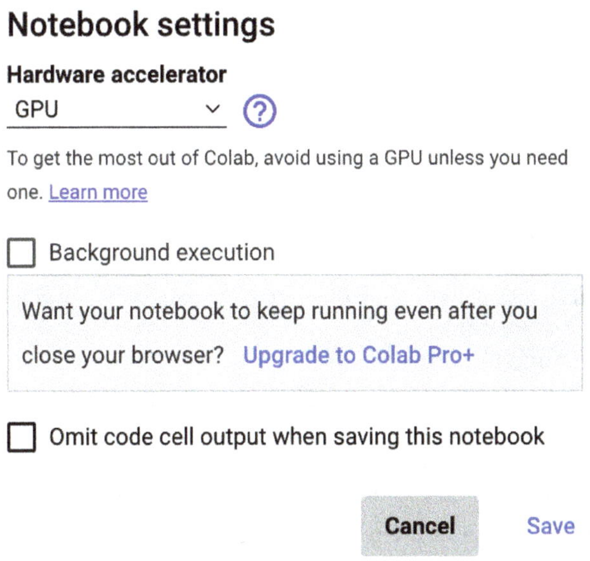

Notebook settings

Hardware accelerator

GPU   ⌄   ⑦

To get the most out of Colab, avoid using a GPU unless you need one. Learn more

☐ Background execution

Want your notebook to keep running even after you

close your browser? Upgrade to Colab Pro+

☐ Omit code cell output when saving this notebook

Cancel   Save

Figure 1.3: Select GPU for Hardware accelerator

Detectron2 has a dedicated tutorial on how to install Detectron2 on Google Colab. However, this section discusses each step and gives further details about these. First, Detectron2 is built on top of PyTorch, so we need to have PyTorch installed. By default, Google Colab runtime already installs PyTorch. So, you can use the following snippet to install Detectron2 on Google Colab:

```
!python -m pip install \
'git+https://github.com/facebookresearch/detectron2.git'
```

If you have an error message such as the following one, it is safe to ignore it and proceed:

```
ERROR: pip's dependency resolver does not currently take into
account all the packages that are installed. This behaviour is
the source of the following dependency conflicts.
flask 1.1.4 requires click<8.0,>=5.1, but you have click 8.1.3
which is incompatible.
```

However, if you face problems such as PyTorch versions on Google Colab, they may not be compatible with Detectron2. Then, you can install Detectron2 for specific versions of PyTorch and CUDA. You can use the following snippet to get PyTorch and CUDA versions:

```
import torch
TORCH_VERSION = ".".join(torch.__version__.split(".")[:2])
CUDA_VERSION = torch.__version__.split("+")[-1]
print("torch: ", TORCH_VERSION, "; cuda: ", CUDA_VERSION)
```

After understanding the PyTorch and CUDA versions, you can use the following snippet to install Detectron2. Please remember to replace TORCH_VERSION and CUDA_VERSION with the values found in the previous snippet:

```
!python -m pip install detectron2 -f \
https://dl.fbaipublicfiles.com/detectron2/wheels/{TORCH_
VERSION}/{CUDA_VERSION}/index.html
```

Here is an example of such an installation command for CUDA version 11.3 and PyTorch version 1.10:

```
!python -m pip install detectron2 -f \
https://dl.fbaipublicfiles.com/detectron2/wheels/cu113/
torch1.10/index.html
```

If you face an error such as the following, it means that there is no matching Detectron2 distribution for the current versions of PyTorch and CUDA:

```
ERROR: Could not find a version that satisfies the requirement
detectron2 (from versions: none)
ERROR: No matching distribution found for detectron2
```

In this case, you can visit the Detectron2 installation page to find the distributions compatible with the current PyTorch and CUDA versions. This page is available at https://detectron2.readthedocs.io/en/latest/tutorials/install.html.

Figure 1.4 shows the current Detectron2 distributions with corresponding CUDA/CPU and PyTorch versions:

Install Pre-Built Detectron2 (Linux only)

Choose from this table to install v0.6 (Oct 2021):

CUDA	torch 1.10	torch 1.9	torch 1.8
11.3	▶ install		
11.1	▶ install	▶ install	▶ install
10.2	▶ install	▶ install	▶ install
10.1			▶ install
cpu	▶ install	▶ install	▶ install

Figure 1.4: Current Detectron2 distributions for corresponding CUDA/CPU and PyTorch versions

Suppose Detectron2 does not have a distribution that matches your current CUDA and PyTorch versions. Then, there are two options. The first option is to select the Detectron2 version with CUDA and PyTorch versions that are closest to the ones that you have. This approach should generally work. Otherwise, you can install the CUDA and PyTorch versions that Detectron2 supports.

Finally, you can use the following snippet to check the installed Detectron2 version:

```
import detectron2
print(detectron2.__version__)
```

Congratulations! You are now ready to develop CV applications using Detectron2 on Google Colab. Read on if you want to create Detectron2 applications on a local machine. Otherwise, you can go to *Chapter 2* to start developing Detectron2 CV applications.

Local development environment for Detectron2 applications

Google Colab is an excellent cloud environment to quickly start building deep learning applications. However, it has several limitations. For instance, the free Google Colab plan may not have enough RAM and GPU resources for large projects. Another limitation is that your runtime may terminate if your kernel is idle for a while. Even in the purchased Pro+ plan, a Google Colab kernel can only run for 24 hours, after which it is terminated. That said, if you have a computer with GPUs, it is better to install Detectron2 on this local computer for development.

> **Important note – resume training option**
>
> Due to time limitations, Google Colab may terminate your runtime before your training completes. Therefore, you should train your models with a resumable option so that the Detectron2 training process can pick up the stored weights from its previous training run. Fortunately, Detectron2 supports a resumable training option so that you can do this easily.

At the time of writing this book, Detectron2 supports Linux and does not officially support Windows. You may refer to its installation page for some workarounds at https://detectron2.readthedocs.io/en/latest/tutorials/install.html if you want to install Detectron2 on Windows. This section covers the steps to install Detectron2 on Linux. Detectron2 is built on top of PyTorch. Therefore, the main installation requirement (besides Python itself) is PyTorch. Please refer to PyTorch's official page at https://pytorch.org/ to perform the installation. *Figure 1.5* shows the interface to select appropriate configurations for your current system and generate a PyTorch installation command at the bottom.

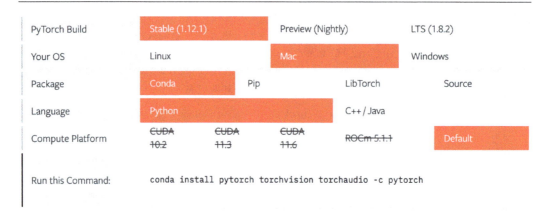

PyTorch Build	Stable (1.12.1)		Preview (Nightly)		LTS (1.8.2)
Your OS	Linux		Mac		Windows
Package	Conda	Pip		LibTorch	Source
Language	Python			C++ / Java	
Compute Platform	CUDA 10.2	CUDA 11.3	CUDA 11.6	ROCm 5.1.1	Default
Run this Command:	conda install pytorch torchvision torchaudio -c pytorch				

Figure 1.5: PyTorch installation command generator (https://pytorch.org)

The next installation requirement is to install Git to install Detectron2 from source. Git is also a tool that any software developer should have. Especially since we are developing relatively complex CV applications, this tool is valuable. You can use the following steps to install and check the installed Git version from the Terminal:

```
$ sudo apt-get update
$ sudo apt-get install git
$ git --version
```

Once PyTorch and Git are installed, the steps to install Detectron2 on a local computer are the same as those used to install Detectron2 on Google Colab, described in the previous section.

Connecting Google Colab to a local development environment

There are cases where developers have developed some code with Google Colab, or they may want to use files stored on Google Drive or prefer to code with the Google Colab interface more than the standard Jupyter notebook on a local computer. In these cases, Google Colab provides an option to execute its notebook in a local environment (or other hosted runtimes such as Google Cloud instances). Google Colab has instructions for this available here: https://research.google.com/colaboratory/local-runtimes.html.

> **Important note – browser-specific settings**
>
> The following steps are for Google Chrome. If you are using Firefox, you must perform custom settings to allow connections from HTTPS domains with standard WebSockets. The instructions are available here: https://research.google.com/colaboratory/local-runtimes.html.

We will first need to install Jupyter on the local computer. The next step is to enable the jupyter_ http_over_ws Jupyter extension using the following snippet:

```
$ pip install jupyter_http_over_ws
$ jupyter serverextension enable --py jupyter_http_over_ws
```

The next step is to start the Jupyter server on the local machine with an option to trust the WebSocket connections so that the Google Colab notebook can connect to the local runtime, using the following snippet:

```
$ jupyter notebook \
--NotebookApp.allow_origin=\
'https://colab.research.google.com' \
--port=8888 \
--NotebookApp.port_retries=0
```

Once the local Jupyter server is running, in the Terminal, there is a backend URL with an authentication token that can be used to access this local runtime from Google Colab. *Figure 1.6* shows the steps to connect the Google Colab notebook to a local runtime: **Connect | Connect to a local runtime**:

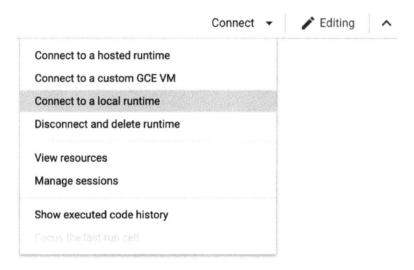

Figure 1.6: Connecting the Google Colab notebook to a local runtime

On the next dialog, enter the backend URL generated in the local Jupyter server and click the **Connect** button. Congratulations! You can now use the Google Colab notebook to code Python applications using a local kernel.

Summary

This chapter discussed advanced CV tasks, including object detection, instance segmentation, keypoint detection, semantic segmentation, and panoptic segmentation, and when to use them. Detectron2 is a framework that helps implement cutting-edge algorithms for these CV tasks with the advantages of being faster, more accurate, modular, customizable, and built on top of PyTorch. Its architecture has four main parts: input data, backbone, region proposal, and region of interest heads. Each of these components is replaceable with a custom implementation. This chapter also provided the steps to set up a cloud development environment using Google Colab, a local development environment, or to connect Google Colab to a local runtime if needed.

You now understand the leading CV tasks Detectron2 can help develop and have set up a development environment. The next chapter (*Chapter 2*) will guide you through the steps to build CV applications for all the listed CV tasks using the cutting-edge models provided in the Detectron2 Model Zoo.

2

Developing Computer Vision Applications Using Existing Detectron2 Models

This chapter guides you through the steps to develop applications for computer vision tasks using state-of-the-art models in Detectron2. Specifically, this chapter discusses the existing and pre-trained models in Detectron2's Model Zoo and the steps to develop applications for object detection, instance segmentation, key-point detection, semantic segmentation, and panoptic segmentation using these models. Thus, you can quickly develop practical computer vision applications without having to train custom models.

By the end of this chapter, you will know what models are available in the Detectron2 Model Zoo. Furthermore, you will learn how to develop applications for vision tasks using these cutting-edge models from Detectron2. Specifically, this chapter covers the following:

- Introduction to Detectron2's Model Zoo
- Developing an object detection application
- Developing an instance segmentation application
- Developing a key-point detection application
- Developing a panoptic segmentation application
- Developing a semantic segmentation application
- Putting it all together

Technical requirements

You should have set up the development environment with the instructions provided in *Chapter 1*. Thus, if you have not done so, please complete setting up the development environment before continuing. All the code, datasets, and respective results are available on the GitHub page of the book (under the folder named Chapter02) at https://github.com/PacktPublishing/Hands-On-Computer-Vision-with-Detectron2. It is highly recommended to download the code and follow along.

Introduction to Detectron2's Model Zoo

In deep learning, when developing large models and training models on massive datasets, developers of the deep learning methods often provide pre-trained models. The main reason for this is that the developers of these models usually are the big players in the field (e.g., Facebook, Google, Microsoft, or universities) with access to computation resources (e.g., CPUs and GPUs) to train such models on large datasets. These computation resources are generally not accessible for standard developers elsewhere.

These pre-trained models are often trained on the datasets for uses with common tasks that the models are intended for. Therefore, another benefit of such models is for the users to adopt them and use them for specific cases of the tasks at hand if they match what the models were trained for. Furthermore, these models can also be used as the baselines on which the end users can further fine-tune them for their specific cases with smaller datasets that require fewer computation resources to train.

Detectron2 also provides an extensive collection of models trained with Big Basin servers with high computation resources (8 NVIDIA V100 GPUs and NVLink) on its Model Zoo. You can read further information and download all the models from this link: https://github.com/facebookresearch/detectron2/blob/main/MODEL_ZOO.md.

It would be a little overwhelming for new users to grasp information about the available models with different specifications. So, this section details which information users should look for to complete a specific computer vision task at hand using these cutting-edge models from Detectron2 Model Zoo. These items include finding information about the following:

- The dataset on which the model was trained
- The architecture of this model
- Training and inferencing times
- Training memory requirement
- Model accuracy
- The link to the trained model (weights)
- The link to its configuration file (.yaml)

Knowing the dataset on which the model was trained gives information about the available tasks that this specific model can offer. For instance, a dataset with class labels and bounding boxes only provides models for object detection tasks. Similarly, datasets that have class labels, bounding boxes, and pixel-level annotations per individual object offer models that provide object detection and instance segmentation, and so on. Furthermore, the specific class labels from the datasets are also significant. For instance, if the computer vision task is detecting humans from videos, and the dataset on which the model was trained does not have a class label for humans (people), then such a model cannot be used for this specific task directly.

The two most common image datasets for computer vision tasks are ImageNet and **COCO (Common Objects in Context)**. ImageNet currently has more than 14 million images (14,197,122 images by this writing). Out of these 14 million images, more than 1 million have bounding boxes for the dominant object in the image. Note also that when the computer vision community refers to ImageNet, they refer to a subset of these images curated by **ImageNet Large Scale Visual Recognition Challenge (ILSVRC)**. This subset has 1,000 object categories and more than 1.2 million images. More than half of these images are annotated with bounding boxes around the objects of the image category. This dataset is often used to train cutting-edge deep learning models such as VGG, Inception, ResNet, and RestNeXt. The number of classes in this dataset is enormous. Therefore, your computer vision task at hand would likely fall to detecting one or more of the 1,000 specified categories. Thus, the chance of reusability of the models trained on this dataset is high.

Microsoft is also another major player in this area. Specifically, Microsoft also publishes the COCO dataset. This dataset contains (at the time of writing this book) 330,000 images, with more than 200,000 images that are labeled. There are about 1.5 million object instances with 80 object categories and 91 stuff categories. Notably, there are also 250,000 people with keypoints to support training models for keypoint detection tasks. All Detectron2's pre-trained COCO models were trained and evaluated on the COCO 2017 dataset. The train, validation, and test sets of COCO 2017 can be downloaded using the following commands:

```
$ wget http://images.cocodataset.org/zips/train2017.zip
$ wget http://images.cocodataset.org/zips/val2017.zip
$ wget http://images.cocodataset.org/zips/test2017.zip
```

> Important note
> ImageNet and COCO datasets are being used and frequently updated by the computer vision community. Interested readers can refer to the official web pages of these datasets for the latest updates and computer vision challenges related to them. These web pages are https://image-net.org and https://cocodataset.org, respectively.

The next important piece of information related to any pre-trained model is the architecture of the pre-trained model. This is often reflected by the model's name in the Model Zoo. For instance, R50 and R101 mean **Microsoft Research Asia** (**MSRA**) ResNet-50 and ResNet-101, respectively. Similarly, X101 and X152 mean ResNeXt-101-32x8d and ResNeXt-152-32x8d, respectively. Some model names include FPN. This part of the name indicates that the model uses the **Feature Pyramid Network** (**FPN**) technique. This technique means that the features are extracted at different layers of the model's architecture (instead of taking the output feature from the last layer and feed to the prediction process). Using the FPN backbone helps to increase the accuracy with the time trade-off. Furthermore, having C4 in its name means that this model uses the conv4 backbone. Similarly, having DC5 in its name indicates that this model uses conv5 with dilations in conv5.

> **Important note**
>
> Interested readers can refer to the web pages regarding the ResNet architecture at https://github.com/KaimingHe/deep-residual-networks and the ResNeXt architecture at https://github.com/facebookresearch/ResNeXt for further information.

After knowing the trained dataset and the model architecture, the next step is to check for the training and inferencing time, training memory, and model accuracy. This information helps to decide the hardware requirements and the trade-off between speed and accuracy. This information is also listed for every pre-trained model on the Detectron2 Model Zoo.

After deciding which model to work with based on speed versus hardware requirement trade-off, the next step is downloading the model's configuration file. On the Model Zoo, the model's name is linked with a .yaml configuration file. **YAML** (short for **Yet Another Markup Language**, or **YAML ain't markup language**) is a human-readable programming language for serializing data. It is commonly used for writing configuration files and supporting automation processes. Finally, there is also a link to the weights of the pre-trained model for downloading.

By this time, you should be able to read and understand important information about the pre-trained models available on the Detectron2 Model Zoo. The next step is using some pre-trained models to perform important computer vision tasks. Developing computer vision applications using Detectron2 typically includes the following steps:

- Selecting a configuration file
- Getting a predictor
- Performing inferences
- Visualizing results

The following sections detail these steps for developing object detection applications.

Developing an object detection application

Object detection generally includes object localization and classification. Specifically, deep learning models for this task predict where objects of interest are in an image by giving the bounding boxes around these objects (localization). The following sections detail the steps to develop an object detection application using Detectron2 pre-trained models.

Getting the configuration file

Various pre-trained models for object detection tasks are available on the Detectron2 Model Zoo. These models are listed under the **COCO Object Detection Baselines** header. Specifically, there are three main categories of these baselines, namely the Faster **Region-based Convolution Neural Network (R-CNN)** subheader, the RetinaNet subheader, and the **Region Proposal Network (RPN)** and Fast R-CNN subheader. *Figure 2.1* shows the available pre-trained models currently listed under the Faster R-CNN subheader.

Faster R-CNN:

Name	lr sched	train time (s/iter)	inference time (s/im)	train mem (GB)	box AP	model id	download
R50-C4	1x	0.551	0.102	4.8	35.7	137257644	model | metrics
R50-DC5	1x	0.380	0.068	5.0	37.3	137847829	model | metrics
R50-FPN	1x	0.210	0.038	3.0	37.9	137257794	model | metrics
R50-C4	3x	0.543	0.104	4.8	38.4	137849393	model | metrics
R50-DC5	3x	0.378	0.070	5.0	39.0	137849425	model | metrics
R50-FPN	3x	0.209	0.038	3.0	40.2	137849458	model | metrics
R101-C4	3x	0.619	0.139	5.9	41.1	138204752	model | metrics
R101-DC5	3x	0.452	0.086	6.1	40.6	138204841	model | metrics
R101-FPN	3x	0.286	0.051	4.1	42.0	137851257	model | metrics
X101-FPN	3x	0.638	0.098	6.7	43.0	139173657	model | metrics

Figure 2.1: Pre-trained Faster R-CNN models for object detection

This list details the model name (which reflects its architecture), train and inference times, memory requirement, accuracy (box **AP** or **average precision**), and links to download the model configuration file and weights.

For demonstration purposes, the application described in this section selects the X101-FPN pre-trained model listed at the end of the list for the object detection task. The model's name indicates that it is the RestNeXt-101-32x8d-FPN network architecture. This name is linked with the model's configuration file (the .yaml file), and the following is the link for this specific model:

```
https://github.com/facebookresearch/detectron2/blob/main/
configs/COCO-Detection/faster_rcnn_X_101_32x8d_FPN_3x.yaml
```

Similarly, there is a link to the pre-trained weights of the model (under the **download** column and the **model** links in *Figure 2.1*). These links can be used to download the corresponding configuration and trained weights for this model. However, this model is currently distributed together with Detectron2's installation. For instance, after installing Detectron2 on Google Colab, you can execute the following commands on a code cell to find out what models are released together with the installation (please note the python version as you may need to change it to reflect the current Python version used):

```
!sudo apt-get install tree
!tree /usr/local/lib/python3.7/dist-packages/detectron2/model_
zoo
```

Figure 2.2 shows a part of the resulting output of this code snippet. The output indicates which models are available for COCO-Detection (models for object detection trained on the COCO dataset).

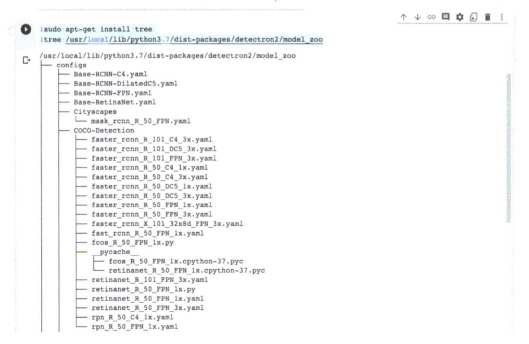

Figure 2.2: Configuration files for models trained on the COCO dataset for object detection

Therefore, only the relative path to this configuration file is necessary. In other words, the configuration file can be accessed from the `configs` folder in the `model_zoo` folder in Detectron2's installation folder. The following is the relative path to the selected configuration file:

```
COCO-Detection/faster_rcnn_X_101_32x8d_FPN_3x.yaml
```

Notably, this is the part at the end of the complete link listed previously after the `/configs/` folder. By reading this configuration file, Detectron2 knows what model configurations are and where to download the model weights. *Figure 2.3* shows the configuration file for the model selected in this section.

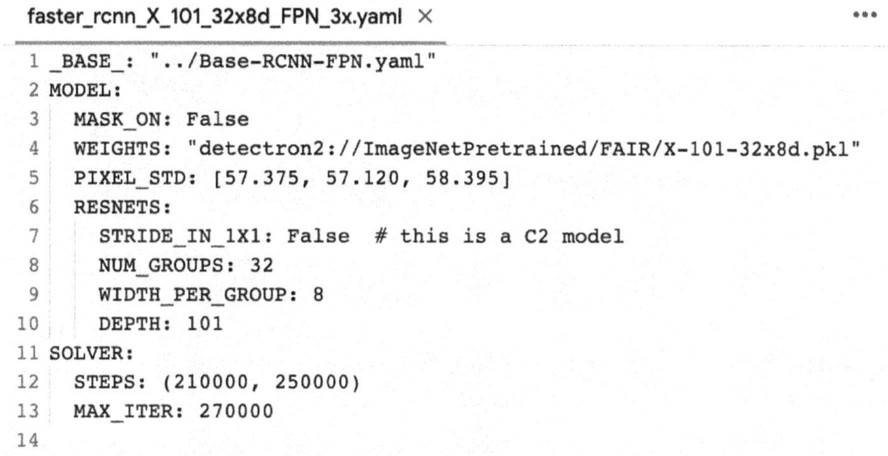

```
faster_rcnn_X_101_32x8d_FPN_3x.yaml  ✕                                      •••

 1 _BASE_: "../Base-RCNN-FPN.yaml"
 2 MODEL:
 3   MASK_ON: False
 4   WEIGHTS: "detectron2://ImageNetPretrained/FAIR/X-101-32x8d.pkl"
 5   PIXEL_STD: [57.375, 57.120, 58.395]
 6   RESNETS:
 7     STRIDE_IN_1X1: False  # this is a C2 model
 8     NUM_GROUPS: 32
 9     WIDTH_PER_GROUP: 8
10     DEPTH: 101
11 SOLVER:
12   STEPS: (210000, 250000)
13   MAX_ITER: 270000
14
```

Figure 2.3: Configuration file for the ResNeXt101 pre-trained model

We now know how to get the configuration file for the intended model. This configuration file also gives information for Detectron2 to download its pre-trained weights. We are ready to use this configuration file and develop the object detection application.

Getting a predictor

The following code snippet imports the required libraries and creates a model (a predictor interchangeably):

```python
import detectron2
from detectron2.config import get_cfg
from detectron2 import model_zoo
from detectron2.engine import DefaultPredictor
# Suppress some user warnings
import warnings
warnings.simplefilter(action='ignore',category=UserWarning)
```

```
# Select a model
config_file = "COCO-Detection/faster_rcnn_X_101_32x8d_FPN_3x.
yaml"
checkpoint_url = "COCO-Detection/faster_rcnn_X_101_32x8d_
FPN_3x.yaml"
# Create a configuration file
cfg = get_cfg()
config_file = model_zoo.get_config_file(config_file)
cfg.merge_from_file(config_file)
# Download weights
cfg.MODEL.WEIGHTS = model_zoo.get_checkpoint_url(checkpoint_
url)
score_thresh_test = 0.95
cfg.MODEL.ROI_HEADS.SCORE_THRESH_TEST = score_thresh_test
predictor = DefaultPredictor(cfg)
```

Specifically, the get_cfg method helps to get a default configuration object for this model. Similarly, model_zoo helps to get the configuration file (the .yaml file) and download the pre-trained weights. At the same time, DefaultPredictor is required to create the predictor from the downloaded configuration file and weights. Besides the required imports, this code snippet also suppresses several user warnings that are safe to ignore.

Notably, the cfg.MODEL.ROI_HEADS.SCORE_THRESH_TEST parameter is currently set to 0.95. This parameter is used at the test time to filter out all the detected objects with detection confidences less than this value. Assigning this parameter to a high value (close to 1.0) means we want to have high precision and low recall. Likewise, setting it to a low value (close to 0.0) means we prefer low precision and high recall. Once we have the model (or predictor interchangeably), it is time to perform inferences.

Performing inferences

The following code snippet performs object detection on an input image. Specifically, we first import cv2 to read an image from an input_url. Next, using the predictor created earlier, it is relatively easy to perform object detection on the loaded image (please upload an image to Google Colab to perform inference):

```
import cv2
input_url = "input.jpeg"
img = cv2.imread(input_url)
output = predictor(img)
```

The next important task is understanding what output format Detectron2 gives and what each output component means. This `print(output)` displays the `output` variable and the fields it provides:

```
{'instances': Instances(num_instances = 4, image_
height = 720, image_width = 960, fields = [pred_boxes:
Boxes(tensor([[492.1558, 331.5353, 687.8397, 637.9542],
    [294.6652, 192.1288, 445.8517, 655.3390],
    [419.8898, 253.9013, 507.7427, 619.1491],
    [451.8833, 395.1392, 558.5139, 671.0604]], device =
'cuda:0')), scores: tensor([0.9991, 0.9986, 0.9945, 0.9889],
device = 'cuda:0'), pred_classes: tensor([0, 0, 0, 0], device =
'cuda:0')])
}
```

Observably, besides the image information, the object detection result also includes the detected instances (`instances`) with corresponding predicted bounding boxes (`pred_boxes`), predicted confidence scores (`scores`), and the predicted class labels (`pred_classes`). With this information, we are ready to visualize the detected objects with corresponding bounding boxes.

Visualizing the results

Detectron2 provides a `Visualizer` class, which helps you to easily annotate detected results to the input image. Thus, the following code snippet imports this class and uses its visualization method to annotate detection results to the input image:

```
from google.colab.patches import cv2_imshow
from detectron2.utils.visualizer import Visualizer
from detectron2.data import MetadataCatalog
metadata = MetadataCatalog.get(cfg.DATASETS.TRAIN[0])
v = Visualizer(img[:, :, ::-1], metadata, scale=0.5)
instances = output["instances"].to("cpu")
annotated_img = v.draw_instance_predictions(instances)
cv2_imshow(annotated_img.get_image()[:, :, ::-1])
```

Specifically, the first statement imports the `cv2_imshow` method, a Google Colab patch, to display images to Google Colab output. The next `import` statement is `Visualizer` itself. The third `import` statement helps get information about the training dataset. Specifically, this `MetadataCatalog` object provides the label names for the predicted objects. In this case, a class label is a `person` for the predicted class number `0`, as in the output.

The `Visualizer` class then uses the original image data and the metadata to create a `Visualizer` object. Notably, this `Visualizer` object uses a different image format than OpenCV (`cv2`). Therefore, we need to use the slice operator as `[:, :, ::-1]` to convert the image format between the two. Finally, we call `draw_instance_predictions` from the visualizer object to annotate the predicted instances to the image and use the `cv2_imshow` method to display the result to the output. *Figure 2.4* shows an example output image after using the Visualizer object to annotate the inference results to the corresponding input image.

Figure 2.4: Inferencing output with object detection results annotated

Congratulations! You have completed the first object detection application using Detectron2. Next, we will move on to develop an instance segmentation application using Detectron2.

Developing an instance segmentation application

Like object detection, instance segmentation also involves object localization and classification. However, instance segmentation takes one step further while localizing the detected objects of interest. Specifically, besides classification, models for this task localize the detected objects at the pixel level. The following sections detail the steps to develop an instance segmentation application using Detectron2 pre-trained models.

Selecting a configuration file

Like object detection, Detectron2 also provides a list of cutting-edge models pre-trained for object instance segmentation tasks. For instance, *Figure 2.5* shows the list of Mask R-CNN models pre-trained on the *COCO Instance Segmentation* dataset.

Name	lr sched	train time (s/iter)	inference time (s/im)	train mem (GB)	box AP	mask AP	model id	download
R50-C4	1x	0.584	0.110	5.2	36.8	32.2	137259246	model | metrics
R50-DC5	1x	0.471	0.076	6.5	38.3	34.2	137260150	model | metrics
R50-FPN	1x	0.261	0.043	3.4	38.6	35.2	137260431	model | metrics
R50-C4	3x	0.575	0.111	5.2	39.8	34.4	137849525	model | metrics
R50-DC5	3x	0.470	0.076	6.5	40.0	35.9	137849551	model | metrics
R50-FPN	3x	0.261	0.043	3.4	41.0	37.2	137849600	model | metrics
R101-C4	3x	0.652	0.145	6.3	42.6	36.7	138363239	model | metrics
R101-DC5	3x	0.545	0.092	7.6	41.9	37.3	138363294	model | metrics
R101-FPN	3x	0.340	0.056	4.6	42.9	38.6	138205316	model | metrics
X101-FPN	3x	0.690	0.103	7.2	44.3	39.5	139653917	model | metrics

Figure 2.5: COCO Instance Segmentation baselines with Mask R-CNN

After checking the specifications of these models, this specific application selects the X101-FPN model. The following is the relative path to this configuration file (extracted and cut out from the address linked with the model name):

```
COCO-InstanceSegmentation/mask_rcnn_X_101_32x8d_FPN_3x.yaml
```

Getting a predictor

The code snippet for getting a predictor for this object instance segmentation remains the same for the previously developed object detection application except for the following two lines. These two lines specify different configuration files for this specific pre-trained model:

```
# Select a model
config_file = "COCO-InstanceSegmentation/mask_rcnn_X_101_32x8d_
FPN_3x.yaml"
checkpoint_url = "COCO-InstanceSegmentation/mask_
rcnn_X_101_32x8d_FPN_3x.yaml"
```

This simple modification is an example of an excellent design of Detectron2. It is relatively easy to switch from one type of application to another by just simply modifying the link to the configuration files.

Performing inferences

The inferencing code snippet remains the same as the object detection application developed in the previous section. However, the object instance segmentation output has some differences. Specifically, besides pred_boxes, scores, and pred_classes, for the bounding boxes, prediction confidence scores, and corresponding predicted class labels, there is also a tensor (pred_masks) to indicate whether an individual pixel from the input picture belongs to a detected instance or not (True/False). The following is an example of this tensor (a part of the output was cut off for space efficiency) for the four detected instances:

```
pred_masks: tensor([
  [[False, False, False,   ..., False, False, False],
   ...,
   [False, False, False,   ..., False, False, False]],

  [[False, False, False,   ..., False, False, False],
   ...,
   [False, False, False,   ..., False, False, False]],

  [[False, False, False,   ..., False, False, False],
   ...,
   [False, False, False,   ..., False, False, False]],

  [[False, False, False,   ..., False, False, False],
   ...,
   [False, False, False,   ..., False, False, False]]],
  device='cuda:0')
```

Visualizing the results

Thanks to the excellent design of Detectron2, visualizing instance segmentation results is the same as visualizing object detection results. In other words, the code snippet remains the same. *Figure 2.6* shows the input image annotated with instance segmentation results. Specifically, besides the bounding boxes and corresponding prediction confidences, pixels belonging to each detected individual are also highlighted.

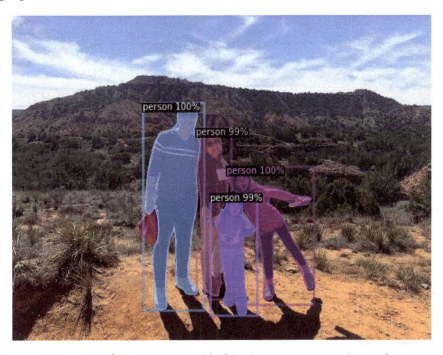

Figure 2.6: Inferencing output with object instance segmentation results

Congratulations! You now completed an object instance segmentation application using a Detectron2 cutting-edge model with only a few modifications compared to the previously developed object detection application. In the next section, you will learn how to develop a keypoint detection application using Detectron2.

Developing a keypoint detection application

 Besides detecting objects, keypoint detection also indicates important parts of the detected objects called **keypoints**. These keypoints describe the detected object's essential trait. This trait is often invariant to image rotation, shrinkage, translation, or distortion. The following sections detail the steps to develop a keypoint detection application using Detectron2 pre-trained models.

Selecting a configuration file

Detectron2 also provides a list of cutting-edge algorithms pre-trained for keypoint detection for human objects. For instance, *Figure 2.7* shows the list of Mask R-CNN models pre-trained on the *COCO Person Keypoint Detection* dataset.

Name	lr sched	train time (s/iter)	inference time (s/im)	train mem (GB)	box AP	kp. AP	model id	download
R50-FPN	1x	0.315	0.072	5.0	53.6	64.0	137261548	model | metrics
R50-FPN	3x	0.316	0.066	5.0	55.4	65.5	137849621	model | metrics
R101-FPN	3x	0.390	0.076	6.1	56.4	66.1	138363331	model | metrics
X101-FPN	3x	0.738	0.121	8.7	57.3	66.0	139686956	model | metrics

Figure 2.7: COCO Person Keypoint Detection baselines with Keypoint R-CNN

In this specific case, we select the X101-FPN pre-trained model. Again, the link to the configuration file is linked with the model name in the first column, and we only use the part after /configs/. Thus, here is the link to the configuration file for this model:

```
COCO-Keypoints/keypoint_rcnn_X_101_32x8d_FPN_3x.yaml
```

Getting a predictor

The code for getting the model remains the same, except for the lines that set the links to the model configuration and weights:

```
# Select a model
config_file = "COCO-Keypoints/keypoint_rcnn_X_101_32x8d_FPN_3x.
yaml"
checkpoint_url = "COCO-Keypoints/keypoint_rcnn_X_101_32x8d_
FPN_3x.yaml"
```

Performing inferences

The code snippet for performing inferences on an input image remains the same. However, the output is a little different this time. Specifically, besides the image information, pred_boxes, scores, and pred_classes, there is a tensor that contains information about the predicted keypoints (pred_keypoints). There is also another field (pred_keypoint_heatmaps) for storing the keypoint heatmaps of the detected person objects. Due to space limitations, this output is not listed here. Please see this application's notebook on the book's GitHub repository, as mentioned in the *Technical requirements* section to view the sample output.

Visualizing the results

Once again, thanks to the flexible design of Detectron2. The visualization code for the keypoint detection application remains the same for object detection and object instance segmentation applications. *Figure 2.8* shows the input image annotated with detected keypoints for detected people.

Figure 2.8: Inferencing output with person keypoint detection results

Congratulations! With only two minor modifications to the object instance segment code, you can now create an application to detect person keypoints using Detectron2. In the next section, you will learn how to develop another computer vision application for panoptic segmentation tasks.

Developing a panoptic segmentation application

Panoptic literally means "everything visible in the image." In other words, it can be viewed as a combination of common computer vision tasks such as instance segmentation and semantic segmentation. It helps to show the unified and global view of segmentation. Generally, it classifies objects in an image into foreground objects (that have proper geometries) and background objects (that do not have appropriate geometries but are textures or materials). The following sections detail the steps to develop a panoptic segmentation application using Detectron2 pre-trained models.

Selecting a configuration file

Detectron2 provides several pre-trained models for panoptic segmentation tasks. For instance, *Figure 2.9* shows the list of panoptic segmentation models trained on the COCO Panoptic Segmentation dataset.

Name	lr sched	train time (s/iter)	inference time (s/im)	train mem (GB)	box AP	mask AP	PQ	model id	download
R50-FPN	1x	0.304	0.053	4.8	37.6	34.7	39.4	139514544	model | metrics
R50-FPN	3x	0.302	0.053	4.8	40.0	36.5	41.5	139514569	model | metrics
R101-FPN	3x	0.392	0.066	6.0	42.4	38.5	43.0	139514519	model | metrics

Figure 2.9: COCO Panoptic Segmentation baselines with panoptic FPN

This specific application in this section selects the R101-FPN model to perform panoptic segmentation tasks. The path to the configuration file of this model is the following:

```
COCO-PanopticSegmentation/panoptic_fpn_R_101_3x.yaml
```

Getting a predictor

Similar to the previous applications, the code snippet for panoptic segmentation remains the same, except for the lines that set the configuration and the weights:

```
# Select a model
config_file = "COCO-PanopticSegmentation/panoptic_fpn_R_101_3x.
yaml"
checkpoint_url = "COCO-PanopticSegmentation/panoptic_
fpn_R_101_3x.yaml"
```

Performing inferences

Once again, the code snippet for performing inferences remains the same. However, the output has some other fields for the panoptic segmentation tasks. Specifically, it contains a field called panoptic_seg for the predicted panoptic segmentation information. The following is the sample output of this field:

```
'panoptic_seg': (tensor([[6, 6, 6, ..., 6, 6, 6],
[6, 6, 6, ..., 6, 6, 6],
[6, 6, 6, ..., 6, 6, 6],
...,
```

```
[9, 9, 9, ..., 9, 9, 9],
[9, 9, 9, ..., 9, 9, 9],
[9, 9, 9, ..., 9, 9, 9]], device = 'cuda:0', dtype = torch.
int32),
[{ 'id': 1, 'isthing': True, 'score': 0.9987247586250305,
'category_id': 0, 'instance_id': 0 },
{ 'id': 2, 'isthing': True, 'score': 0.997495174407959,
'category_id': 0, 'instance_id': 1 },
{ 'id': 3, 'isthing': True, 'score': 0.9887707233428955,
'category_id': 0, 'instance_id': 2 },
{ 'id': 4, 'isthing': True, 'score': 0.9777324199676514,
'category_id': 0, 'instance_id': 3 },
{ 'id': 5, 'isthing': False, 'category_id': 37, 'area': 15173
},
{ 'id': 6, 'isthing': False, 'category_id': 40, 'area': 163644
},
{ 'id': 7, 'isthing': False, 'category_id': 45, 'area': 192040
},
{ 'id': 8, 'isthing': False, 'category_id': 46, 'area': 62182
},
{ 'id': 9, 'isthing': False, 'category_id': 47, 'area': 164592
}])
```

Specifically, this field is a tuple with two elements. The first element is a tensor of the same size as the input image. This tensor helps to indicate the detected class label number for every pixel from the input image. The second element of this tuple is a list of items that map corresponding label numbers into known classes.

Visualizing the results

The visualization code snippet for panoptic segmentation is very similar to other vision tasks we have learned, except that we need to extract the two elements of the panoptic_seg field (described previously) from the output. Next, these two fields are passed to the draw_panoptic_seg_predictions method of the Visualizer object to perform the annotations:

```
from google.colab.patches import cv2_imshow
from detectron2.utils.visualizer import Visualizer
from detectron2.data import MetadataCatalog
metadata = MetadataCatalog.get(cfg.DATASETS.TRAIN[0])
v = Visualizer(img[:, :, ::-1], metadata, scale=0.5)
```

```
panoptic_seg, segments_info = output["panoptic_seg"]
annotated_img = v.draw_panoptic_seg_predictions(panoptic_seg.
to("cpu"), segments_info)
cv2_imshow(annotated_img.get_image()[:, :, ::-1])
```

Figure 2.10 shows the panoptic segmentation results for an input image. Specifically, every pixel in the input image is classified as either one of the four detected people, the sky, a mountain, grass, a tree, or dirt (a total of nine label values).

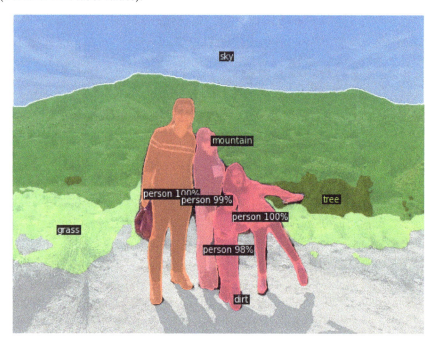

Figure 2.10: Inferencing output with person keypoint detection results

Congratulations! You now know how to develop a cutting-edge panoptic segmentation application. Though there are a few changes for this panoptic segmentation application compared to the previous applications, the main code is pretty much the same. In the next section, you will learn how to develop an application for semantic segmentation tasks.

Developing a semantic segmentation application

A semantic segmentation task does not detect specific instances of objects but classifies each pixel in an image into some classes of interest. For instance, a model for this task classifies regions of images into pedestrians, roads, cars, trees, buildings, and the sky in a self-driving car application. The following sections detail the steps to develop a semantic segmentation application using Detectron2 pre-trained models.

Selecting a configuration file and getting a predictor

Semantic segmentation is a byproduct of panoptic segmentation. For instance, it groups detected objects of the same class into one if they are in the same region instead of providing segmentation data for every detected object. Therefore, the model for semantic segmentation is the same as that for panoptic segmentation. Therefore, the configuration file, the weights, and the code snippet for getting a predictor are the same as those for the previous panoptic segmentation application.

Performing inferences

The inferencing code snippet for semantic segmentation is also the same for panoptic segmentation. The inferencing result (the same as that for panoptic segmentation) also has a field called `sem_seg`. This field has the shape of `classes` × `image_height` × `image_width`. Specifically, it is 54 × 720 × 960 in this specific case. The size 720 × 960 corresponds to all pixels of the input image. For each pixel, there are 54 values, corresponding to the probabilities of the pixel being classified into 54 categories of the trained dataset, respectively.

Visualizing the results

The visualization of the semantic segmentation results is slightly different. First, the `sem_seg` field is extracted. Next, from the 54 probabilities, the class label with the highest predicted probability is selected as the label for every individual pixel using the `argmax` method. The extracted data is then passed to the `draw_sem_seg` method from the `Visualizer` object to perform the visualization. Here is the code snippet for visualizing the semantic segmentation result:

```
from google.colab.patches import cv2_imshow
from detectron2.utils.visualizer import Visualizer
from detectron2.data import MetadataCatalog
metadata = MetadataCatalog.get(cfg.DATASETS.TRAIN[0])
v = Visualizer(img[:, :, ::-1], metadata, scale=0.5)
sem_seg = output["sem_seg"].argmax(dim=0)
annotated_img = v.draw_sem_seg(sem_seg.to("cpu"))
cv2_imshow(annotated_img.get_image()[:, :, ::-1])
```

Figure 2.11 shows the semantic segmentation result visualized on the corresponding input image. Notably, compared to the panoptic results, the detected people are grouped into one region and classified as `things`.

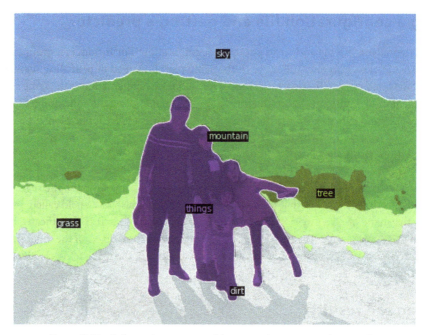

Figure 2.11 – Inferencing output with semantic segmentation results

Congratulations! You now mastered the steps to create computer vision applications to perform important computer vision tasks using cutting-edge models available on Detectron2. Thanks to Detectron2's flexible design, the code snippets for these applications are very similar to one another. Therefore, the next section puts all these together and develops an application that can perform all these tasks.

Putting it all together

After developing different applications for different computer vision tasks in the previous sections, it is clear that Detectron2 is highly structured because the described applications have the same structure. Therefore, it is reasonable to implement these steps to perform all these tasks. The following sections cover the steps to develop a computer vision application that can perform different computer vision tasks using Detectron2 pre-trained models.

Getting a predictor

The code snippets for getting a predictor for different computer vision tasks are similar. They all need to perform some basic imports, take the configuration file, weight file, and score threshold at test time. Therefore, we create a function called get_predictor for this. Here is the code snippet for this function:

```
import detectron2
from detectron2.config import import get_cfg
```

```
from detectron2 import model_zoo
from detectron2.engine import DefaultPredictor
# We suppress some user warnings
import warnings
warnings.simplefilter(action='ignore',category=UserWarning)
# Create a configuration and a predictor
def get_predictor(config_file, checkpoint_url, score_thresh_
test=1.0):
  # create a configuration object
  cfg = get_cfg()
  # get the configurations from the config_file
  config_file = model_zoo.get_config_file(config_file)
  cfg.merge_from_file(config_file)
  # get the pre-built weights of a trained model from the
checkpoint
  cfg.MODEL.WEIGHTS = model_zoo.get_checkpoint_url(checkpoint_
url)
  # set the threshold for recall vs. precision at test time
  cfg.MODEL.ROI_HEADS.SCORE_THRESH_TEST = score_thresh_test
  # create a predictor
  predictor = DefaultPredictor(cfg)
  return cfg, predictor
```

To get a model, you only need to select the paths to the configuration file and pre-trained weights and pass them to this method. This method then returns a predictor (`predictor`) together with this model's configuration (`cfg`). After having the predictor, the next step is to perform inferences.

Performing inferences

The code snippet for all the computer vision tasks is precisely the same. Therefore, we should provide a method for this (`perform_inference`). Here is the code snippet for performing inference given any predictor (`predictor`) and the path to the input image (`input_url`):

```
import cv2
from google.colab.patches import cv2_imshow
def perform_inference(predictor, input_url):
  img = cv2.imread(input_url)
  output = predictor(img)
  return img, output
```

This method (`perform_inference`) returns the input image and corresponding output predicted by the predictor on the input image. The next step is to have a method to annotate this output to the input image.

Visualizing the results

The code snippets for visualizing object detection, object instance segmentation, and keypoint detection are the same. However, they are slightly different if the task is panoptic or semantic segmentations. Therefore, there are constants to indicate which task to be performed, and the visualization method (called `visualize_output`) can perform appropriate visualization based on the input task. Here is the code snippet for the result visualizations:

```python
from detectron2.utils.visualizer import Visualizer
from detectron2.data import MetadataCatalog
# computer vision tasks
OBJECT_DETECTION = 0
INSTANCE_SEGMENTATION = 1
KEYPOINT_DETECTION = 2
SEMANTIC_SEGMENTATION = 3
PANOPTIC_SEGMENTATION = 4

def visualize_output(img, output, cfg, task=OBJECT_DETECTION,
scale=1.0):
  v = Visualizer(img[:, :, ::-1], MetadataCatalog.get(cfg.
DATASETS.TRAIN[0]), scale=scale)
  if task == PANOPTIC_SEGMENTATION:
    panoptic_seg, segments_info = output["panoptic_seg"]
    annotated_img = v.draw_panoptic_seg_predictions(panoptic_
seg.to("cpu"), segments_info)
  elif task == SEMANTIC_SEGMENTATION:
    sem_seg = output["sem_seg"].argmax(dim=0)
    annotated_img = v.draw_sem_seg(sem_seg.to("cpu"))
  else:
    annotated_img = v.draw_instance_
predictions(output["instances"].to("cpu"))
  cv2_imshow(annotated_img.get_image()[:, :, ::-1])
```

Performing a computer vision task

After having all the corresponding methods for getting a predictor (get_predictor), performing inference (perform_inference), and visualizing output (visualize_output), we are ready to perform detections. Here is the code snippet for performing object detection tasks using these methods:

```
config_file = "COCO-Detection/faster_rcnn_X_101_32x8d_FPN_3x.
yaml"
checkpoint_url = "COCO-Detection/faster_rcnn_X_101_32x8d_
FPN_3x.yaml"
score_thresh_test = 0.95
input_url = "input.jpeg"
task = OBJECT_DETECTION
scale = 0.5
# get a predictor
cfg, predictor = get_predictor(config_file, checkpoint_url,
score_thresh_test)
# perform inference
img, output = perform_inference(predictor, input_url)
# visualization
visualize_output(img, output, cfg, task, scale)
```

Other computer vision tasks can be performed in the same way. Furthermore, these methods can be placed into a library (a *.py file), so you can include them and use them in any computer vision applications if appropriate.

Summary

This chapter discussed the pre-trained models available on the Detectron2 Model Zoo. Specifically, it specified the vital information to look for while selecting a pre-trained model for a computer vision task. It then provided the code snippets for developing an object detection application, an object instance segmentation application, a keypoint detection application, a panoptic segmentation application, and a semantic segmentation application. The code snippets for these applications are similar. Therefore, the code snippets are abstracted into typical methods for getting a model, performing inferences, and visualizing outputs. These methods can then be reused to develop different computer vision applications with Detectron2 quickly.

The cutting-edge models on Detectron2 are trained with many images and a vast amount of computation resources. Therefore, these models have high accuracies. Additionally, they are also trained on the datasets with class labels for everyday computer vision tasks. Thus, there is a high chance that they can meet your individual needs. However, if Detectron2 does not have a model that meets your requirement, you can train models on custom datasets using Detectron2. *Part 2* of this book covers the steps to prepare data, train, and fine-tune a Detectron2 model on a custom dataset for the object detection task.

Part 2: Developing Custom Object Detection Models

The second part is about getting your hands busy developing and fine-tuning custom detection models with data preparation, training, and fine-tuning steps. The data preparation step introduces common computer vision datasets and the code to download freely available images. Additionally, it discusses tools to label data, common annotation formats, and the code to convert from different formats to the one Detectron2 supports. It then goes into further detail about the architecture of a Detetron2 application using visualizations and code. After training a model, this part illustrates the steps to utilize TensorBoard to find insights about training before fine-tuning the trained models. For fine-tuning, this section provides a primer on deep-learning optimizers and steps to fine-tune Detectron2 solvers. For optimizing detection models specifically, this part includes the code to compute the suitable sizes and ratios parameters for generating anchors for Detectron2 models. Finally, this part provides a tour of augmentation techniques and Detectron2's system for performing augmentations at training and testing.

The second part covers the following chapters:

- *Chapter 3, Data Preparation for Object Detection Applications*
- *Chapter 4, The Architecture of the Object Detection Model in Detectron2*
- *Chapter 5, Training Custom Object Detection Models*
- *Chapter 6, Inspecting Training Results and Fine-Tuning Detectron2's Solver*
- *Chapter 7, Fine-Tuning Object Detection Models*
- *Chapter 8, Image Data Augmentation Techniques*
- *Chapter 9, Applying Train-Time and Test-Time Image Augmentations*

3

Data Preparation for Object Detection Applications

This chapter discusses the steps to prepare data for training models using Detectron2. Specifically, it provides tools to label images if you have some datasets at hand. Otherwise, it points you to places with open datasets so that you can quickly download and build custom applications for computer vision tasks. Additionally, this chapter covers the techniques to convert standard annotation formats to the data format required by Detectron2 if the existing datasets come in different formats.

By the end of this chapter, you will know how to label data for object detection tasks and how to download existing datasets and convert data of different formats to the format supported by Detectron2. Specifically, this chapter covers the following:

- Common data sources
- Getting images
- Selecting image labeling tools
- Annotation formats
- Labeling the images
- Annotation format conversions

Technical requirements

You should have set up the development environment with the instructions provided in *Chapter 1*. Thus, if you have not done so, please complete setting up the development environment before continuing. All the code, datasets, and results are available on the GitHub page of the book (under the folder named Chapter03) at https://github.com/PacktPublishing/Hands-On-Computer-Vision-with-Detectron2. It is highly recommended to download the code and follow along.

Common data sources

Chapter 2 introduced the two most common datasets for the computer vision community. They are **ImageNet** and Microsoft **COCO (Common Objects in Context)**. These datasets also contain many pre-trained models that can predict various class labels that may meet your everyday needs.

If your task is to detect a less common class label, it might be worth exploring the **Large Vocabulary Instance Segmentation** (**LVIS**) dataset. It has more than 1,200 categories and 164,000 images, and it contains many rare categories and about 2 million high-quality instance segmentation masks. Detectron2 also provides pre-trained models for predicting these 1,200+ labels. Thus, you can follow the steps described in *Chapter 2* to create a computer vision application that meets your needs. More information about the LVIS dataset is available on its website at https://www.lvisdataset.org.

If you have a task where no existing/pre-trained models can meet your needs, it is time to find existing datasets and train your model. One of the common places to look for datasets is **Kaggle** (https://www.kaggle.com). Kaggle is a Google subsidiary. It provides an online platform for data scientists and machine learning practitioners to share datasets, train machine learning models, and share the trained models and their thoughts via discussions. It also often holds and allows users to host data science and machine learning competitions (including computer vision competitions, of course). At the time of writing, searching the phrase "object detection" on Kaggle brings 555 datasets, as shown in *Figure 3.1*.

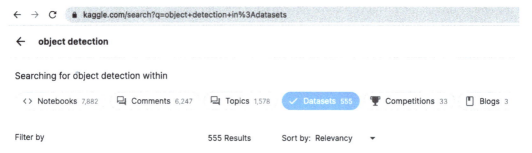

Figure 3.1: Datasets on Kaggle for object detection tasks

Kaggle contains datasets for data science and machine learning in general, including computer vision datasets. However, if you still cannot find a dataset for your custom needs, you may be interested in exploring **Roboflow Universe** available at https://universe.roboflow.com. It claims to be the world's most extensive collection of open source computer vision datasets and APIs. At the time of writing, more than 100 million images, 110,000 datasets, and 10,000 pre-trained models are available on Roboflow Universe, as shown in *Figure 3.2*. Datasets for common computer vision tasks are available, such as object detection, instance segmentation, and semantic segmentation.

Figure 3.2: Datasets on Roboflow Universe

Computer vision is advancing rapidly, and there is a vast number of available practitioners in this field. If you cannot find the datasets that meet your needs in the preceding two resources, you can use the internet (e.g., Google) to search for the dataset that you want. However, if you still cannot find such a dataset, it is time to collect and label your own dataset. The following sections cover the standard tools and steps for collecting and labeling datasets for computer vision applications.

Getting images

When you start a new project, you might have a set of images collected using graphic devices such as cameras, smartphones, drones, or other specialized devices. If you have some photos, you can skip this step and move to the following steps, where you select a labeling tool and start labeling. However, if you are about to start a fresh new project and do not even have any images, you can get some ideas from this section on collecting pictures from the internet. There might be copyrights related to images that you collect from the internet. Thus, you should check for copyrights to see whether the downloaded images are usable for your purposes.

Google Images contains a massive number of photos, which continues to grow every day. Therefore, it is a great resource to crawl images for training your custom model. Python has a `simple_image_download` package that provides scripts to search for images using keywords or tags and download them. First, we will install this package using the following command:

```
!pip install simple_image_download
```

You can run this command in Google Colab's code cell to install it and download images to a running Google Colab instance. However, once the instance is stopped, the downloaded photos are gone. Therefore, you should download the images to a mapped Google Drive folder or on a local computer. You can run the same command on a locally hosted Jupyter notebook or remove the start (!) character and execute it in a terminal to install this package locally. If you are running on Google Colab, the following code snippet helps to map Google Drive and creates a folder in Google Drive:

```
import os
from google.colab import drive
drive.mount('/gdrive')
project_folder = "/gdrive/My Drive/Detectron2/Chapter3"
os.makedirs(project_folder, exist_ok=True)
```

The next code snippet (to be executed in a separate Google Colab code cell) sets the working directory to the previously created folder, so the downloaded results remain in this folder when the Google Colab instance stops:

```
cd "/gdrive/My Drive/Detectron2/Chapter3"
```

It is always a good practice to ensure that Google Colab is currently working in this directory by running the pwd() command in another code cell and inspecting the output.

Once you are in the specified folder, the following code snippet helps to download images from Google Images using the simple_image_download package:

```
from tqdm import tqdm
from simple_image_download.simple_image_download import simple_
image_download as Downloader
def _download(keyword, limit):
  downloader = Downloader()
  downloader.download(keywords=keyword, limit=limit)
  urls = downloader.urls(keywords=keyword, limit=limit)
  return urls

def download(keywords, limit):
  for keyword in tqdm(keywords):
    urls = _download(keyword=keyword, limit=limit)
    with open(f"simple_images/{keyword}.txt", "w") as f:
      f.writelines('\n'.join(urls))
```

Specifically, the first import statement helps to import the tqdm package, which can be used to show downloads' progress. This is important because downloading a large number of images from the internet would take time. Therefore, it would be helpful to have a progress bar showing us that it is currently working and approximately how long it will take to complete the task. The next import statement imports the simple_image_download object, which helps to perform the downloading task. The next is the _download helper method to download a number (specified by the limit parameter) of images from Google using a keyword (specified by the keyword parameter). Furthermore, for the corresponding pictures, this helper method also returns an array of URLs that can be stored to validate the downloaded images and check for copyrights if needed.

The next method is called `download`. It uses the previous helper to download images, and the results are stored under a folder called `simple_images`. The images for each keyword are stored in a subfolder named after the keyword itself. Furthermore, this method writes a text file per keyword for all the corresponding URLs to the downloaded images with the same name as the keyword plus the `.txt` extension. For instance, the following code snippet downloads 10 images for the `brain tumors x-ray` and `heart tumors x-ray` keywords:

```
download(keywords=["brain tumors x-ray", "heart tumors x-ray"],
limit=10)
```

Figure 3.3 shows the resulting folders in Google Drive for the downloaded images and text files to store the image URLs on the internet. These URLs can be used to validate the images or check for copyrights.

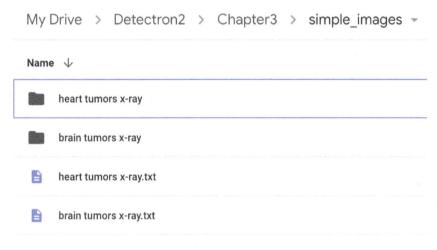

Figure 3.3: Resulting image folders and text files for their corresponding URLs

The downloaded images may not be what you are looking for exactly. Therefore, you should check the downloaded images and delete those that do not meet your specific needs. Once you have the desired images, the next step is to select a tool to start labeling the images for training and validation.

Selecting an image labeling tool

Computer vision applications are developing rapidly. Therefore, there are many tools for labeling images for computer vision applications. These tools range from free and open source to fully commercial or commercial with free trials (which means there are some limitations regarding available features). The labeling tools may be desktop applications (require installations), online web applications, or locally hosted web applications. The online applications may even provide cloud storage, utilities to support collaborative team labeling, and pre-trained models that help to label quicker (generally, with some cost).

One of the popular image labeling tools is `labelImg`. It is available at `https://github.com/heartexlabs/labelImg/blob/master/README.rst`. It is an open source, lightweight, fast, easy-to-use, Python-based application with a short learning curve.

Its limitation is that it supports only the creation of rectangle bounding boxes. It is currently part of **Label Studio** (which is a more complicated tool with many more features). However, it is still the best choice if the task at hand is to perform object detection and localization only.

Another popular tool is `labelme` (`https://github.com/wkentaro/labelme`). It is similar to `labelImg` in terms of being a Python-based application, open source, fast, and easy to use. However, it supports other annotation types besides rectangle bounding boxes. Specifically, it can annotate polygons, rectangles, circles, lines, and points. Therefore, it can be used to label training images for object detection and semantic or instance segmentation tasks.

Both `labelImg` and `labelme` are locally installed applications. However, if you want to work with a locally hosted web application, you may refer to **VGG** (**Visual Geometry Group**) **Image Annotator** (**VIA**). VIA is a lightweight, standalone, browser-based web application. It also supports annotating polygons, circles, lines, and rectangular bounding boxes.

If you want to use a cloud-based web application (that supports cloud storage and annotation on the web), you may like `roboflow`, available at `https://app.roboflow.com`. This online web application supports various annotation modes. It has utilities for teams to label images collaboratively. It also supports pre-trained models to help label images faster. However, several of its advanced features are only available in the paid plan. Another popular web platform for this task is `https://www.makesense.ai`. It is open source and free to use under a GPLv3 license. It supports several popular annotation formats and uses AI to make your labeling job more productive.

This chapter selects `labelImg` to prepare data for the object detection tasks. *Chapter 10* uses `labelme` to prepare data for the object instance segmentation tasks. After choosing a labeling tool, the next step is to discover which annotation formats that tool supports, which one to select, and how to convert it to the format that Detectron2 supports.

Annotation formats

Similar to labeling tools, many different annotation formats are available for annotating images for computer vision applications. The common standards include COCO JSON, Pascal VOC XML, and YOLO PyTorch TXT. There are many more formats (e.g., TensorFlow TFRecord, CreateML JSON, and so on). However, this section covers only the previously listed three most common annotation standards due to space limitations. Furthermore, this section uses two images and labels extracted from the test set of the brain tumor object detection dataset available from Kaggle (`https://www.kaggle.com/datasets/davidbroberts/brain-tumor-object-detection-datasets`) to illustrate these data formats and demonstrate their differences, as shown in *Figure 3.4*. This section briefly discusses the key points of each annotation format, and interested readers can refer to the GitHub page of this chapter to inspect this same dataset in different formats in further detail.

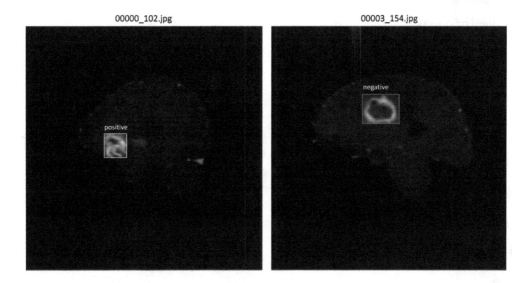

Figure 3.4: Two images and tumor labels used to illustrate different annotation formats

The COCO **JSON (JavaScript Object Notation)** format is famous thanks to the popularity of the Microsoft COCO dataset. The Detectron2 data format is based on COCO's annotation standard. Therefore, it is worth diving into this dataset a little bit deeper. Each dataset in COCO's annotation format is represented in one JSON file. This JSON file is stored in the same folder with all the images. Thus, it is often named with "_" as the start character to bring it to the top of the list view when viewing this data folder. COCO's annotation JSON file contains these main elements:

- The info section (`info`) provides general descriptions of the dataset

- The licenses section (`licenses`) is a list of licenses applied to the images

- The categories section (`categories`) is a list of the available categories (or class labels) and supercategories for this dataset

- The images section (`images`) is a list of image elements, and each has information such as the `id`, `width`, `height`, and `file_name` of the image

- The annotations section (`annotations`) is a list of annotations, and each annotation has information such as segmentation and a bounding box (`bbox`)

The following snippet provides sample sections of a dataset with two images in the test set (available from `https://www.kaggle.com/datasets/davidbroberts/brain-tumor-object-detection-datasets`) in COCO annotation format:

Here is the info section:

```
"info": {
        "year": "2022",
        "version": "1",
        "description": "Brain tumor object detection",
        "contributor": "",
        "url": "<the URL to Kaggle dataset>",
        "date_created": "2022-10-07T13:54:09+00:00"
    },
```

This is the licenses section:

```
"licenses": [
        {
            "id": 1,
            "url": "https://creativecommons.org/publicdomain/
zero/1.0/",
            "name": "CC0 1.0"
        }
    ],
```

This is the categories section:

```
"categories": [
        {
            "id": 0,
            "name": "tumors",
            "supercategory": "none"
        },
        {
            "id": 1,
            "name": "negative",
            "supercategory": "tumors"
        },
        {
            "id": 2,
            "name": "positive",
            "supercategory": "tumors"
```

```
        }
    ],
```

Here is the images section:

```
    "images": [
        {
            "id": 0,
            "license": 1,
            "file_name": "00000_102.jpg",
            "height": 512,
            "width": 512,
            "date_captured": "2022-10-07T13:54:09+00:00"
        },
        {
            "id": 1,
            "license": 1,
            "file_name": "00003_154.jpg",
            "height": 512,
            "width": 512,
            "date_captured": "2022-10-07T13:54:09+00:00"
        }
    ],
```

Here is the annotations section:

```
    "annotations": [
        {
            "id": 0,
            "image_id": 0,
            "category_id": 2,
            "bbox": [170,226,49,51],
            "area": 2499,
            "segmentation": [],
            "iscrowd": 0
        },
        {
            "id": 1,
```

```
        "image_id": 1,
        "category_id": 1,
        "bbox": [197,143,80,62],
        "area": 4960,
        "segmentation": [],
        "iscrowd": 0
    }
]
```

Notably, the bounding box is specified in absolute pixel locations. The coordinate system has the height as the *y* axis, the width as the *x* axis, and the origin at the top-left corner of the picture. For instance, for the first image (00003_154.jpg and image_id = 1), the bounding box is [197,143,80,62], which are the values for [x_min, y_min, width, height], respectively, as shown in *Figure 3.5*.

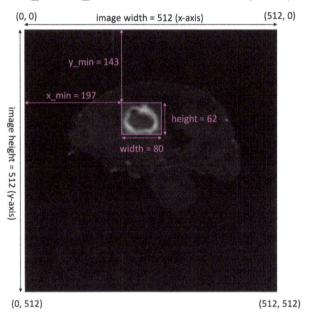

Figure 3.5: Coordinate system and annotation format ([x_min, y_min, width, height]) in COCO style for bounding box

Another popular image annotation format is the **Pascal VOC (Pattern Analysis, Statistical Modeling, and Computational Learning Visual Object Classes)** format. This annotation format is in the **XML (eXtensible Markup Language)** file format, and there is one XML file per image.

The annotation filename has the same name as the image filename with the `.xml` extension instead. Storing one XML file per image prevents having one large file for all the annotations for a large image dataset. The following is the annotation file (`00003_154.xml`) for one image in the demo dataset (`00003_154.jpg`):

```
<annotation>
    <folder></folder>
    <filename>00003_154.jpg</filename>
    <path>00003_154.jpg</path>
    <source>
        <database>annotationdemo</database>
    </source>
    <size>
        <width>512</width>
        <height>512</height>
        <depth>3</depth>
    </size>
    <segmented>0</segmented>
    <object>
        <name>negative</name>
        <pose>Unspecified</pose>
        <truncated>0</truncated>
        <difficult>0</difficult>
        <occluded>0</occluded>
        <bndbox>
            <xmin>197</xmin>
            <xmax>277</xmax>
            <ymin>143</ymin>
            <ymax>205</ymax>
        </bndbox>
    </object>
</annotation>
```

Observably, this annotation file contains image information and annotation objects. As shown in *Figure 3.6*, the image coordinate system is the same as described previously for the COCO annotation format. It also has four numbers for the bounding box: `xmin`, `xmax`, `ymin`, and `ymax`. In other words, the bounding box is stored using the top-left and bottom-right coordinates in absolute pixel values (instead of the top-left corner and width and height values).

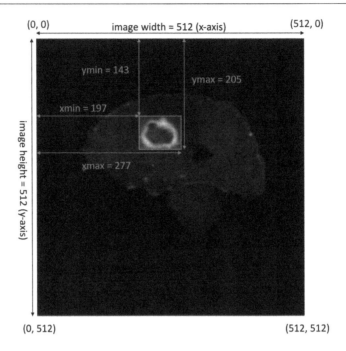

Figure 3.6: Coordinate system and annotation format ([xmin, xmax, ymin, ymax]) in Pascal VOC style for bounding box

Due to the recent development of **YOLO (You Only Look Once)**, YOLO data annotation is another popular data format. Similar to the PASCAL VOC annotation style, the dataset has one annotation file per image. However, the annotation file extension is .txt instead.

A dataset is organized into two subfolders, one for images (images) and one for annotations (labels). These subfolders might be further divided into the train (train), test (test), and validation (val) subfolders. There is also a text file called classes.txt (or possibly a *.yaml file) that declares the classes or labels for this dataset, one per line. The following is the content of the classes.txt file for our simple demo dataset with two classes:

```
negative
positive
```

Different from the COCO and Pascal VOC annotation styles, YOLO stores one annotation per line. Specifically, there are five values in a line per bounding box. They are the label index (`label_idx`), the center coordinates (`center_x, center_y`), and then the width (`width`) and height (`height`) of the bounding box. The label index is the map of the label into its corresponding index (starting from 0 for the first label). For instance, in this specific case, `negative` is mapped into 0, and `positive` is mapped into 1.

For example, the annotation for the `00003_154.jpg` image is `00003_154.txt`, and it has the following content:

```
0 0.463028 0.339789 0.156103 0.120892
```

These values correspond to [`label_idx, center_x, center_y, width, height`], as described, for one bounding box. Notably, the coordinate of the center, width, and height of the bounding box are all converted into values as ratios of the corresponding values to the image size. *Figure 3.7* illustrates this annotation style for the `00003_154.jpg` image.

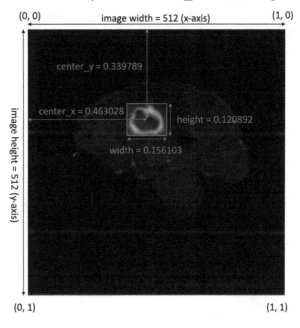

Figure 3.7: Coordinate system and annotation format ([center_x, center_y, width, height]) in YOLO style for bounding box

Congratulations! At this point, you should clearly understand the essential and popular annotation formats used by the computer vision community to annotate images. The next step is to label a dataset for your custom computer vision task. The following section guides you through the steps that use `labelImg` to prepare data for the object detection task. *Chapter 10* uses `labelme` to prepare data for the object instance segmentation tasks.

Labeling the images

This chapter uses `labelImg` to perform data labeling for object detection tasks. This tool requires installation on the local computer. Therefore, if you downloaded your images using Google Colab to your Google Drive, you need to map or download these images to a local computer to perform labeling. Run the following snippet in a terminal on a local computer to install `labelImg` using Python 3 (if you are running Python 2, please refer to the `labelImg` website for a guide):

```
pip3 install labelImg
```

After installing, run the following snippet to start the tool, where `[IMAGE_PATH]` is an optional argument to specify the path to the image folder, and `[PRE-DEFINED CLASS FILE]` is another optional argument to indicate the path to a text file (`*.txt`) that defines a list of class labels (one per line):

```
labelImg [IMAGE_PATH] [PRE-DEFINED CLASS FILE]
```

For instance, after downloading/synchronizing the `simple_images` folder for the downloaded images in the previous steps, we can set the working directory of the terminal to this directory and use the following command to start labeling the pictures on the `"brain tumors x-ray"` folder:

```
labelImg "brain tumors x-ray"
```

Figure 3.8 shows the main interface of `labelImg` with the main menu on the left, the list of pictures on the bottom right, the current labeling image at the center, and the list of labels on this current image in the top-right corner.

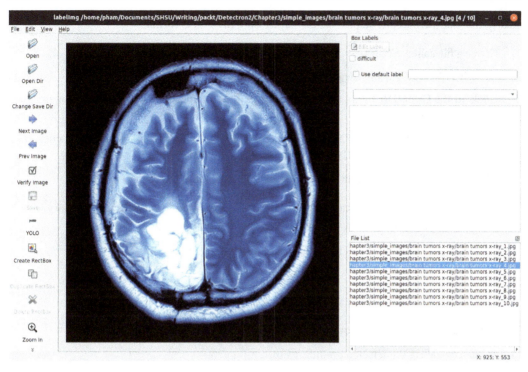

Figure 3.8: The labellmg interface for bounding box labeling

This tool supports exporting the data annotations into standard data formats such as Pascal VOC and YOLO. After installing and launching the application (using the preceding instructions), the following are the steps to label images using the Pascal VOC format:

1. Click on the **Open Dir** menu to open the image directory (if you did not specify the image directory at launching time).

2. Click on the **Change Save Dir** menu to change the annotation directory.

3. Click on **Create RectBox** to start annotating rectangular boxes.

4. Click and release the left mouse button to select a region to annotate the bounding box. You can use the right mouse button to drag the rectangle box to copy or move it.

> **Important note**
>
> Please refer to the previous section regarding the folder structure of YOLO to create corresponding folders for images and labels before starting labeling using the following steps.

Similarly, the following are the steps to label images using the YOLO format:

1. Create a file called `classes.txt` with a list of labels (one label per line).

2. Launch `labelImg` using the previous instructions.

3. Click on the **Open Dir** menu to open the image directory (if you did not specify the image directory at launching time).

4. If the current format is not YOLO (it might currently be CreateML or Pascal VOC), click on the **CreateML** or **Pascal VOC** menu until it switches to **YOLO**.

5. Click on the **Change Save Dir** menu to change the annotation directory.

A `.txt` file per image with annotation in the YOLO format is saved in the selected directory with the same name. A file named `classes.txt` is saved in that folder too.

Additionally, the following shortcuts help support quick labeling:

- *The W* key: Create a bounding box
- *The D* key: Move to the next image
- *The A* key: Move to the previous image
- *Delete* key: delete the selected bounding box
- Arrow keys: Move the selected bounding box
- *Ctrl + Shift + D*: Delete the current image

Acquiring datasets from different sources or data labeling tools may come with annotation formats that differ from what Detectron2 supports (the COCO data format). Therefore, knowing how to convert datasets from different formats to the COCO format is essential. The following section guides you through performing this dataset conversion process.

Annotation format conversions

Detectron2 has its data description built based on the COCO annotation format. In other words, it supports registering datasets using the COCO data annotation format. However, other data annotation formats are abundant, and you may download a dataset or use a labeling tool that supports another data format different from COCO. Therefore, this section covers the code snippets used to convert data from the popular Pascal VOC and YOLO formats to COCO style.

> **Important note**
> A statement that starts with an exclamation mark (`!`) means it is a Bash command to be executed in a Jupyter notebook (Google Colab) code cell. If you want to run it in a terminal, you can safely remove this exclamation mark and execute this statement.

By understanding the different data formats as described, it is relatively easy to write code to convert data from one format to another. However, this section uses the pylabel package to perform this conversion to speed up the development time. In a Jupyter notebook or Google Colab code cell, run the following command to install the pylabel package:

```
!pip install pylabel
```

Converting YOLO datasets to COCO datasets

This section also uses the simple annotation demo dataset described previously to demonstrate how these conversions work. Therefore, you can upload the YOLO and Pascal VOC datasets to Google Colab (if you are using Google Colab) or set the working directory of your local Jupyter notebook instance to the dataset folder (annotationdemo). The following is the code snippet for downloading the yolo.zip file, which stores the dataset in YOLO format into the current Google Colab running instance and unzips it:

```
!wget url_to_yolo.zip
!unzip yolo.zip
```

Optionally, you can install the tree command and list this simple dataset to view its structure using this snippet:

```
!sudo apt-get install tree
!tree yolo
```

The following is the sample output of this statement for this dataset:

```
yolo
├ classes.txt
├ images
|    └ test
|          ├ 00000_102.jpg
|          └ 00003_154.jpg
└ labels
     └ test
           ├ 00000_102.txt
           └ 00003_154.txt
```

The following snippet helps create a new COCO format dataset structure:

```
import os
import shutil
from glob import glob
from tqdm import tqdm


annotations_path = "yolo/labels/test"
images_path = "yolo/images/test"
coco_dir = 'coco/test'
os.makedirs(coco_dir, exist_ok=True)


txt_files = glob(os.path.join(annotations_path, "*.txt"))
img_files = glob(os.path.join(images_path, "*.jpg"))
# copy annotations
for f in tqdm(txt_files):
  shutil.copy(f, coco_dir)
# copy images
for f in tqdm(img_files):
  shutil.copy(f, coco_dir)
```

Specifically, COCO datasets typically have images placed in a folder. Therefore, we create a folder and place images into this folder. Additionally, this snippet copies the YOLO annotation files to this folder to perform the conversion. Once the conversion is completed, these files are removed.

The following snippet reads the classes from the YOLO dataset (`classes.txt`) and imports the YOLO annotations copied to the COCO dataset directory previously (`coco_dir`) to a `pylabel` dataset:

```
from pylabel import importer
# get the classes
with open("yolo/classes.txt", "r") as f:
  classes = f.read().split("\n")
# load dataset
dataset = importer.ImportYoloV5(path=coco_dir, cat_
names=classes, name="brain tumors")
```

Notably, at the time of writing, YOLO v7 is available; however, it has the same annotation format as YOLO v5, and data produced with this annotation format is popular. Thus, it is safe to import the dataset as YOLO v5. After importing the dataset, the following code snippet exports the dataset to COCO format and stores it in the _annotations.coco.json file:

```
# export
coco_file = os.path.join(coco_dir, "_annotations.coco.json")
# Detectron requires starting index from 1
dataset.export.ExportToCoco(coco_file, cat_id_index=1)
```

Notably, Detectron2 requires the label index to start with index 1. Thus, ExportToCoco sets cat_id_index = 1. After performing the conversion, the YOLO annotation files under coco_dir should be removed using this snippet:

```
# now delete yolo annotations in the coco set
for f in txt_files:
    os.remove(f.replace(annotations_path, coco_dir))
```

Now, the conversion is completed. You can use the following snippet to view the structure of the generated dataset in COCO format:

```
!tree coco
```

The following is the output of this converted dataset:

```
coco
└ test
    ├ 00000_102.jpg
    ├ 00003_154.jpg
    └ _annotations.coco.json
```

Congratulations! By this time, you should have your dataset in COCO format, and it should be ready to be used by Detectron2. However, if your dataset is in Pascal VOC annotation format, the following section helps to perform this conversion.

Converting Pascal VOC datasets to COCO datasets

The Pascal VOC annotation format is another popular one. So, this section provides code snippets to convert it to the COCO annotation format before registering it with Detectron2. This section uses the simple dataset described previously as a demo (annotationdemo). First, we download the brain tumors dataset in Pascal VOC (voc.zip) and extract it on Google Colab:

```
# Download the dataset and unzip it
!wget url_to_voc.zip
!unzip voc.zip
```

With the dataset extracted, the following code snippet installs the tree package and views the structure of this dataset:

```
!sudo apt-get install tree
!tree voc
```

The following is the structure of the dataset in COCO annotation format:

```
voc
└ test
     ├ 00000_102.jpg
     ├ 00000_102.xml
     ├ 00003_154.jpg
     └ 00003_154.xml
```

Notably, the Pascal VOC data format stores one annotation file per image file with the same name and in a .xml extension.

This dataset can be modified to be in COCO format in place. However, the following code snippet copies these images and annotations into a new directory (coco_dir) for the COCO dataset and avoids modifying the original one:

```
import os
import shutil
from glob import glob
from tqdm import tqdm

voc_dir = "voc/test"
coco_dir = 'coco/test'
os.makedirs(coco_dir, exist_ok=True)
```

```
xml_files = glob(os.path.join(voc_dir, "*.xml"))
img_files = glob(os.path.join(voc_dir, "*.jpg"))
# copy annotations
for f in tqdm(xml_files):
  shutil.copy(f, coco_dir)
# copy images
for f in tqdm(img_files):
  shutil.copy(f, coco_dir)
```

Once copied, it is relatively easy to load this dataset as a `pylabel` dataset using the following code snippet:

```
from pylabel import importer
# load dataset
dataset = importer.ImportVOC(coco_dir, name="brain tumors")
```

Similarly, it is relatively easy to export the loaded dataset to a COCO style and store it in the `_annotations.coco.json` file with the following code snippet:

```
# export
coco_file = os.path.join(coco_dir, "_annotations.coco.json")
# Detectron requires starting index from 1
dataset.export.ExportToCoco(coco_file, cat_id_index=1)
```

Notably, Detectron2 requires the label index (`cat_id_index`) to start from 1. Therefore, the `ExportToCoco` method sets this value to 1.

Once the conversion is completed, it is safe to delete the Pascal VOC annotation files (`.xml` files) in the COCO dataset (`coco_dir`) using the following code snippet:

```
# now delete yolo annotations in the coco set
for f in xml_files:
  os.remove(f.replace(voc_dir, coco_dir))
```

Optionally, you can use the following statement to view the structure of the resulting dataset:

```
!tree coco
```

The following is the structure of the dataset in COCO annotation format:

```
coco
└ test
    ├ 00000_102.jpg
    ├ 00003_154.jpg
    └ _annotations.coco.json
```

Congratulations! By now, you should be able to convert datasets in Pascal VOC into COCO datasets that are ready to be consumed by Detectron2. If you have datasets in different formats, the steps to perform conversions should be similar.

Summary

This chapter discussed the popular data sources for the computer vision community. These data sources often have pre-trained models that help you quickly build computer vision applications. We also learned about the common places to download computer vision datasets. If no datasets exist for a specific computer vision task, this chapter also helped you get images by downloading them from the internet and select a tool for labeling the downloaded images. Furthermore, the computer vision field is developing rapidly, and many different annotation formats are available. Therefore, this chapter also covered popular data formats and the steps to convert these formats into the format supported by Detectron2.

By this time, you should have your dataset ready. The next chapter discusses the architecture of Detectron2 with details regarding the backbone networks and how to select one for an object detection task before training an object detection model using Detectron2.

The Architecture of the Object Detection Model in Detectron2

This chapter dives deep into the architecture of Detectron2 for the object detection task. The object detection model in Detectron2 is the implementation of Faster R-CNN. Specifically, this architecture includes the backbone network, the region proposal network, and the region of interest heads. This chapter is essential for understanding common terminology when designing deep neural networks for vision systems. Deep understanding helps to fine-tune and customize models for better accuracy while training with the custom datasets.

By the end of this chapter, you will understand Detectron2's typical architecture in detail. You also know where to customize your Detectron2 model (what configuration parameters to set, how to set them, and where to add/remove layers) to improve performance. Specifically, this chapter covers the following topics:

- Introduction to the application architecture
- The backbone network
- The region proposal network
- The region of interest heads

Technical requirements

You should have set up the development environment with the instructions provided in *Chapter 1*. Thus, if you still need to do so, please complete setting up the development environment before continuing. Additionally, you should read *Chapter 2* to understand Detectron2's Model Zoo and what backbone networks are available. All the code, datasets, and results are available on the GitHub page of the book (in the folder named `Chapter04`) at `https://github.com/PacktPublishing/Hands-On-Computer-Vision-with-Detectron2`. It is highly recommended that you download the code and follow along.

Introduction to the application architecture

As discussed in *Chapter 1* and shown in *Figure 4.1*, Detectron2 has the architecture with the backbone network, the region proposal network, and the region of interest heads.

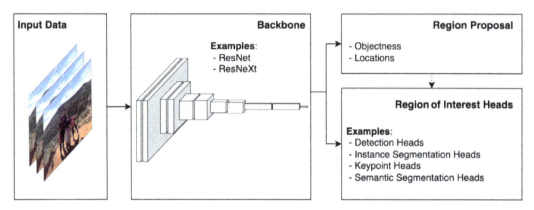

Figure 4.1: The main components of Detectron2

The backbone network includes several convolutional layers that help to perform feature extraction from the input image. The region proposal network is another neural network that predicts the proposals with objectness and locations of the objects before feeding to the next stage. The region of interest heads have neural networks for object localization and classification. However, the implementation details of Detectron2 are more involved. We should understand this architecture in depth to know what Detectron2 configurations to set and how to fine-tune its model.

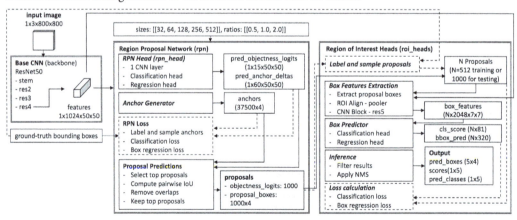

Figure 4.2: Architecture of Detectron2's implementation of Faster R-CNN

Detectron2's architecture for object detection tasks is the implementation of Faster **Region-based Convolution Neural Network (R-CNN)**. The architecture of Faster R-CNN is depicted in *Figure 4.2* in detail. The styles of the boxes indicate their meaning. Specifically, shaded boxes are data (input, intermediate results, and output), dashed boxes are tasks performed during training only, and double-line boxes are the neural networks.

The following sections discuss these individual items, how they work together, and where to configure them using Detectron2. These sections use an example model for object detection tasks, as in *Chapter 2*. All the following tasks can be performed by calling the predict method from the default predictor. However, this chapter walks through each step with input, code, and intermediate results to inspect the steps and components of this model. Checking individual components with code helps explain why each step is essential, what parameters to change in each step, and where to customize the model. First, we need to import the torch package and install Detectron2:

```
import torch
print(torch.__version__)
!python -m pip install \
'git+https://github.com/facebookresearch/detectron2.git'
```

Next, we will import Detectron2, get configuration and model checkpoint files, and create a predictor. Specifically, we use ResNet50 (R50-C4) in this example because it is simple to illustrate, and its configuration file is COCO-Detection/faster_rcnn_R_50_C4_1x.yaml. However, the idea is the same for other backbone networks:

```
import detectron2
from detectron2.config import get_cfg
from detectron2 import model_zoo
from detectron2.engine import DefaultPredictor
# Suppress some user warnings
import warnings
warnings.simplefilter(action='ignore', category=UserWarning)
# Select a model
config_file = "COCO-Detection/faster_rcnn_R_50_C4_1x.yaml"
checkpoint_url = "COCO-Detection/faster_rcnn_R_50_C4_1x.yaml"
# Create a configuration file
cfg = get_cfg()
config_file = model_zoo.get_config_file(config_file)
cfg.merge_from_file(config_file)
# Download weights
cfg.MODEL.WEIGHTS = model_zoo.get_checkpoint_url(checkpoint_
```

```
url)
score_thresh_test = 0.95
cfg.MODEL.ROI_HEADS.SCORE_THRESH_TEST = score_thresh_test
predictor = DefaultPredictor(cfg)
```

One input image of size 800×800 (this is the default input size, so we do not need to do a post-processing step for resizing the outputs) is passed through different parts of this architecture to inspect what each component does. The following snippet helps to prepare the input:

```
!wget https://raw.githubusercontent.com/phamvanvung/hcvd2/
main/800x800/input.jpeg
import cv2
input_url = "input.jpeg"
img = cv2.imread(input_url)
height, width = img.shape[:2]
image = torch.as_tensor(img.astype("float32").transpose(2, 0,
1))
```

Specifically, this snippet first uses `cv2` to load the input image and get its size (height and width). Additionally, it converts the loaded image into a PyTorch tensor with the channel-first format. Specifically, `cv2` loads images in the (`height, width, channel`) form, or the (`H, W, C`) format, and Detectron2 works with (`channel, height, width`) or (`C, H, W`).

Once the default model and the input are ready, it is time to pass the prepared input through different components of the model and inspect what each part does.

The backbone network

Chapter 2 discusses the typical backbone networks for Detectron2. They include ResNet50, ResNet101, ResNeXt101, and their variants. This section inspects the ResNet50 architecture as an example. However, the idea remains the same for other base models (backbone networks). *Figure 4.3* summarizes the steps to inspect the backbone network. Specifically, we pass a tensor of data for a single image to the backbone, and the backbone (ResNet50, in this case) gives out a tensor. This output tensor is the extracted salient feature of the input image.

Figure 4.3: The backbone network

Specifically, from the default Detectron2's predictor, we can access the backbone network using the following code snippet:

```
backbone = predictor.model.backbone
type(backbone)
```

This code snippet should print out the following:

```
detectron2.modeling.backbone.resnet.ResNet
```

The following code snippet reveals the backbone's architecture:

```
print(backbone)
```

The output is long, so it is not placed here for space efficiency. Generally, it has different convolutional layers organized in four main blocks (stem, res2, res3, and res4). The typical principle is that after passing through each block, the features become smaller, and the depth becomes bigger to extract more abstract or salient features. The following code snippet passes the input image through the blocks of this backbone and tracks the output of each block in this network:

```
with torch.no_grad():
    rnInput = image.to("cuda").unsqueeze(0)
    stem = backbone.stem(rnInput)
    res2 = backbone.res2(stem)
    res3 = backbone.res3(res2)
    res4 = backbone.res4(res3)
    print(rnInput.shape)
```

```
print(stem.shape)
print(res2.shape)
print(res3.shape)
print(res4.shape)
```

Specifically, this snippet converts the input tensor into `"cuda"` to load that image to GPU because our model currently runs on GPU. The `unsqueeze` method converts the image input into the (NxCxHxW) format, where N is the number of images, and we pass one image in this case. In other words, this step results in a tensor of size `1x3x800x800` as the input to this backbone. This input is then fed to the main blocks of ResNet50 (`stem`, `res2`, `res3`, and `res4`), and the whole code blocks should have the following output:

```
torch.Size([1, 3, 800, 800])
torch.Size([1, 64, 200, 200])
torch.Size([1, 256, 200, 200])
torch.Size([1, 512, 100, 100])
torch.Size([1, 1024, 50, 50])
```

Figure 4.4 visualizes these main blocks and their corresponding output sizes in the ResNet50 backbone. Notably, the further the blocks go, the depth increases, and the size decreases to extract more abstract or salient features from the image.

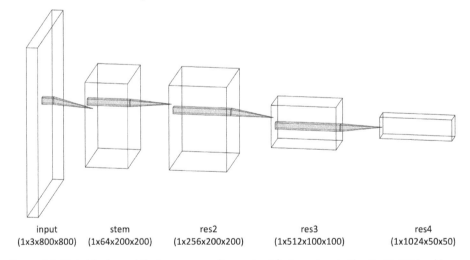

Figure 4.4: Main blocks and their corresponding output feature sizes in the ResNet50 backbone

The output of this backbone (extracted features or feature maps) is the last block called `'res4'` of size `1x1024x50x50`. Conceptually, each point in this feature map corresponds to and is responsible for predicting and classifying bounding boxes related to a region in the original images. In other words, these extracted features correspond to `50×50` `(2,500)` regions in the original image. The following code snippets illustrate this idea. First, we prepare a method to display images in their original sizes:

```python
import numpy as np
import matplotlib.pyplot as plt


def imshow(image):
    dpi = plt.rcParams["figure.dpi"]
    height, width = image.shape[:2]
    figsize = width / float(dpi), height / float(dpi)
    fig = plt.figure(figsize=figsize)
    plt.imshow(image)
    plt.imshow(image)
    plt.axis("off")
    plt.show()
```

We are now ready to draw the center points for the centers of 2,500 regions that correspond to the 50x50 extracted features:

```python
feature_h, feature_w = res4.shape[2:]
sub_sample = height/feature_h
new_img = img.copy()
# the feature map points and corresponding regions
rect_w = 4
rect_h = 4
for i in range(feature_h):
    for j in range(feature_w):
        center_x = sub_sample * j + sub_sample/2
        center_y = sub_sample * i + sub_sample/2
        start_x = int(center_x - rect_w/2)
        start_y = int(center_y - rect_h/2)
        end_x = int(center_x + rect_w/2)
        end_y = int(center_y + rect_h/2)
```

```
    cv2.rectangle(new_img, (start_x, start_y), (end_x, end_y),
(0, 0, 255))
imshow(new_img[:,:,::-1])
```

Specifically, each point in the feature map is projected back to the corresponding point (and region) in the original image, as shown in *Figure 4.5*.

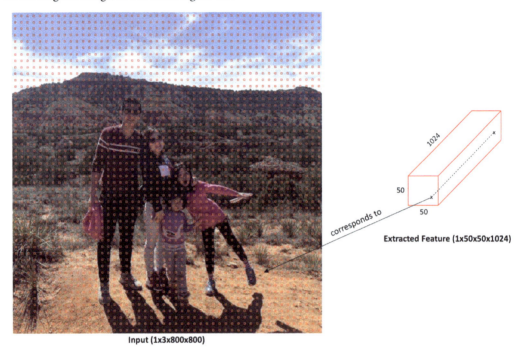

Figure 4.5: Feature maps and their corresponding points (regions) in the original image

For each of these projected points, there are a fixed number of raw anchors (bounding boxes with different sizes and ratios) generated as the initial boxes for performing bounding box proposals using the **Region Proposal Network (RPN)**. The following section details the components of the RPN.

Region Proposal Network

Faster R-CNN is called a two-stage technique. The first stage proposes the regions (bounding boxes) and whether an object falls within that region (objectness). Notably, at this stage, it only predicts whether an object is in the proposed box and does not classify it into a specific class. The second stage then continues to fine-tune the proposed regions and classify objects in the proposed bounding boxes into particular labels. The RPN performs the first stage. This section inspects the details of the RPN and its related components in Faster R-CNN architecture, implemented in Detectron2, as in *Figure 4.6*.

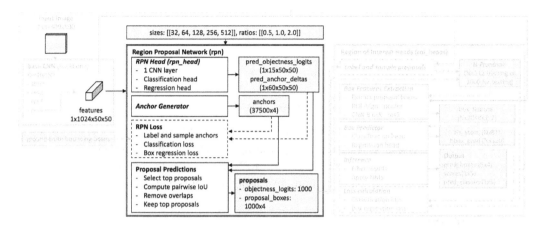

Figure 4.6: The Region Proposal Network and its components

Continuing from the previous code example, the following code snippet displays the RPN (proposal_generator):

```
rpn = predictor.model.proposal_generator
type(rpn)
```

This snippet should print out the following:

```
detectron2.modeling.proposal_generator.rpn.RPN
```

The following statement prints this module to inspect its architecture:

```
print(rpn)
```

This statement should display the following output:

```
RPN(
  (rpn_head): StandardRPNHead(
    (conv): Conv2d(
      1024, 1024, kernel_size=(3, 3), stride=(1, 1),
padding=(1, 1)
      (activation): ReLU()
    )
    (objectness_logits): Conv2d(1024, 15, kernel_size=(1, 1),
stride=(1, 1))
    (anchor_deltas): Conv2d(1024, 60, kernel_size=(1, 1),
stride=(1, 1))
```

```
    )
    (anchor_generator): DefaultAnchorGenerator(
      (cell_anchors): BufferList()
    )
  )
```

Observably, this RPN has two main parts: the standard RPN head (`rpn_head`) and the anchor generator (`anchor_generator`). Additionally, during training, there is a component for RPN to calculate the loss of this RPN network. Finally, this RPN head makes the predictions to propose regions for the next stage. The following sections dive deep into these components.

The anchor generator

The anchor generator is used at the training time to generate all the raw anchors, one set (e.g., 15 anchors or a different number of anchors depending on the configuration, more about this later) per feature pixel (50x50x15 = 37,500 anchors, in this case). Using anchors is a great technique of Faster R-CNN, and other techniques (such as YOLO and SSD) also adopt them to improve prediction accuracies. Specifically, instead of having to predict bounding boxes from "nothing," we can give the model a set of initial bounding boxes (with specific sizes and width/height ratios) and ask the models to predict the bounding boxes relative to these initial boxes (location/bounding-box deltas). Different sizes and ratios help to deal with objects of different scales and shapes.

This anchor generator takes inputs as anchors' ratios and sizes. These two parameters can be set for Detectron2 using configuration variables. The following snippet displays these currently set values:

```
print(cfg.MODEL.ANCHOR_GENERATOR.SIZES)
print(cfg.MODEL.ANCHOR_GENERATOR.ASPECT_RATIOS)
```

This snippet should give the following output:

```
[[32, 64, 128, 256, 512]]
[[0.5, 1.0, 2.0]]
```

Precisely, the sizes correspond to the square root of the areas of the initial anchors in the original image. Additionally, the ratios specify these initial anchors' height/width ratios. In this specific case, *the combinations of these 2 sets of values produce 15 anchors (5 sizes times 3 ratios)*. The different sizes and ratios address the objects of different sizes and aspect ratios. This approach (pyramids of anchors) is considered more cost-efficient compared to other approaches, such as feature pyramids (extracting features at different convolution layers to address different sizes) or using different filter sizes and ratios (sizes of sliding windows).

Recall that the backbone produces an output of size `1x1024x50x50`. This `50x50` feature map size corresponds to 2,500 points in the original image, as shown in *Figure 4.5*. For each point, the region generator generates 15 anchors (5 sizes x 3 ratios). The following code snippets visualize these 15 anchors imposed on top of the original image for the middle point of the feature map:

```
sizes = cfg.MODEL.ANCHOR_GENERATOR.SIZES
ratios = cfg.MODEL.ANCHOR_GENERATOR.ASPECT_RATIOS
new_img = img.copy()
# the feature
for size in sizes[0]:
  for ratio in ratios[0]:
    anchor_h = size*np.sqrt(ratio)
    anchor_w = size/np.sqrt(ratio)
    center_x = sub_sample * feature_h/2 + sub_sample/2
    center_y = sub_sample * feature_w/2 + sub_sample/2
    x0 = int(center_x - anchor_w/2)
    y0 = int(center_y - anchor_h/2)
    x1 = int(center_x + anchor_w/2)
    y1 = int(center_y + anchor_h/2)
    print(x1 - x0, y1 - y0)
    cv2.rectangle(new_img, (x0, y0), (x1, y1), (0, 0, 255))
imshow(new_img[:,:,::-1])
```

Figure 4.7 depicts the result of this code snippet. Observably, there are 15 anchors with respective sizes and ratios for this specific point.

Figure 4.7: The anchors (15) for the center point in the feature map

These anchors are centered in one pixel. Thus, they are a little hard to differentiate. *Figure 4.8* separates them into three groups of the same ratios and five different sizes for illustration purposes.

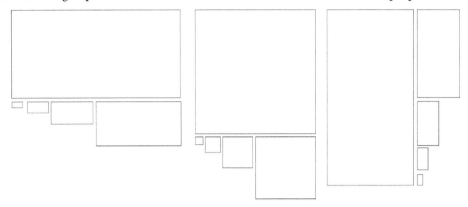

Figure 4.8: The 15 anchors (3 ratios and 5 sizes) centered at 1 feature pixel

Back to Detectron2's implementation, the following code snippet helps inspect how Detectron2 generates these anchors:

```
# generate anchors
anchors = rpn.anchor_generator([res4])
print(anchors[0].tensor.shape)
```

The anchor generator takes the input as the extracted features to calculate the sub_sample value (ratio between the input size and the feature size) and generate anchors. In this case, there are 50×50×15=37,500 anchor boxes. Therefore, the output of this code snippet is the following:

```
torch.Size([37500, 4])
```

Notably, there are 37,500 anchors, each with 4 location specifications (x_0, y_0, x_1, y_1). To confirm, the following code snippet takes the 15 anchors for the middle point of the image and displays it to the original input (as performed previously):

```
new_img = img.copy()
# the feature
col_idx = 24
row_idx = 24
startBox = (row_idx*50 + col_idx)*15
for i in range(startBox, startBox + 15):
    anchors_np = anchors[0].tensor[i].to("cpu").numpy()
    x0, y0, x1, y1 = [int(v) for v in anchors_np]
```

```
     cv2.rectangle(new_img, (x0, y0), (x1, y1), (0, 0, 255))
  imshow(new_img[:,:,::-1])
```

Specifically, the middle point of the `50x50` feature size has row and column indices (`row_idx` and `col_idx`) of `24` (start counting from 0), and each point in the feature map corresponds to 15 anchors. Thus, the start box of the 15 boxes for this middle point is computed as `(row_idx*50 + col_idx)*15`. Next, the snippet takes the coordinates of these boxes and displays them on the image. The output of this code snippet looks the same as shown in *Figure 4.6*.

By this point, the `anchor_generator` component should be clear. The following section inspects the standard RPN head (`rpn_head`).

The RPN head

Executing `rpn.rpn_head` in a code cell displays the components of the RPN head as the following:

```
StandardRPNHead(
   (conv): Conv2d(1024, 1024, kernel_size=(3, 3), stride=(1, 1),
padding=(1, 1) (activation): ReLU())
   (objectness_logits): Conv2d(1024, 15, kernel_size=(1, 1),
stride=(1, 1))
   (anchor_deltas): Conv2d(1024, 60, kernel_size=(1, 1),
stride=(1, 1))
)
```

Observably, this head has one convolutional layer, one classification head, and one regression head. Specifically, the convolution layer (`conv`) performs further salient feature extraction. The classification head predicts whether the input anchor has an object in it or not. It is stored in the variable called `objectness_logits`. Ideally, the output of this component should be probability values ranging from 0 to 1. However, the values in probabilities are small. Thus, the objectness scores are stored as logits to avoid numerical errors.

The `anchor_deltas` head is in charge of predicting how much each of the anchors' coordinates should be adjusted to get the proposals' locations. The following code snippet passes the features extracted from the backbone to this head and generates output:

```
pred_objectness_logits, pred_anchor_deltas = rpn.rpn_head(res4)
print(pred_objectness_logits[0].shape)
print(pred_anchor_deltas[0].shape)
```

This code snippet has the following output:

```
torch.Size([1, 15, 50, 50])
torch.Size([1, 60, 50, 50])
```

Specifically, there are `1x15x50x50` anchors. Thus, there are `1x15x50x50` values in `pred_objectness_logits` to indicate whether there is an object in each of the anchors. Similarly, each of the 15 anchors per feature point has 4 values for coordinates (x_0, y_0, x_1, y_1), which explains the 60 (15x4) for the output of `pred_anchor_deltas`. After having the anchors and related predictions, the next step during the training stage is calculating the RPN loss.

The RPN loss calculation

This loss includes the classification loss from the classification head (predicting objectness scores for anchors) and the regression loss (predicting anchor deltas). Therefore, there must be a label for each of the generated anchors. An anchor is either positive (there is an object, and the label value is `1`), negative (there is no object, and the label value is `0`), or something else (to be ignored, and the label value is `-1`). The label assignment task utilizes the measurement of how much an anchor is overlapped with the ground-truth boxes. The popular measurement for this purpose is **Intersection over Union** (**IoU**). *Figure 4.9* illustrates the computation of IoU.

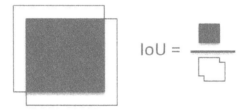

Figure 4.9: Computation of Intersection over Union (IoU)

Specifically, IoU between two boxes is computed by the ratio of the intersection between them and the union between them.

An anchor has a positive label if it (*i*) has the highest IoU with a ground-truth box or (*ii*) has an IoU value that is higher than a specified threshold (e.g., `0.7`) with any ground-truth boxes. The second condition should be sufficient, but the first condition helps in cases where there are rare bounding boxes. An anchor has a negative label if all its IoU values with the ground-truth boxes are smaller than a threshold (e.g., `0.3`). All other labels are ignored (have a label value of `-1`). These two IoU thresholds can be customized using Detectron2 configuration parameters. The following code snippet displays these two values in the current model:

```
print(cfg.MODEL.RPN.IOU_THRESHOLDS)
```

This statement should display [0.3, 0.7] as the two default thresholds. These thresholds can be customized to improve accuracy.

With this labeling strategy, many anchors can be assigned as positively related to one ground-truth bounding box. Similarly, since there is a massive number of anchors (37,500 in this specific case), many of them are also negative. This number of negative anchors dominates the loss calculation and makes the resulting models prefer to predict negative labels for all the bounding boxes. Therefore, for each image, there is a mini-batch of anchors sampled from these anchors, and only these anchors contribute to the loss calculation. The sampling strategy for this mini-batch is as follows:

1. Read the batch size for this mini-batch (anchors per image, as n).
2. Read the fraction of positive anchors in this mini-batch (pos_frac).
3. Compute the number of positive anchors (n_pos = int(n*pos_frac)).
4. Sample the positive anchors:

    I. If the generated raw anchors have more positive anchors than n_pos, sample for n_pos anchors.

    II. Otherwise, set n_pos as the number of all positive anchors.

5. Sample for n_neg = n - n_pos negative anchors.
6. Set all other anchors to a label value of -1 (ignored).

The following code snippet displays the configuration parameters for the mini-batch size and the positive anchor fraction:

```
print(cfg.MODEL.RPN.BATCH_SIZE_PER_IMAGE)
print(cfg.MODEL.RPN.POSITIVE_FRACTION)
```

This code snippet should display 256 as the mini-batch size and 0.5 for the positive anchor fractions. Specifically, for each image, 256 anchors participate in the training (contribute to the loss calculation), of which about half are positive and the other half are negative. All other anchors are ignored (do not contribute to the loss).

This loss calculation step requires the ground-truth boxes. Therefore, the following code snippet prepares a list of ground-truth labels for our list of input images (one image in this case):

```
from detectron2.modeling.roi_heads.roi_heads import Boxes
boxes = Boxes(torch.tensor([[190, 206, 352, 744],
        [419, 375, 633, 711],
        [375, 444, 479, 747],
        [330, 285, 423, 690]], device='cuda:0'))
classes = torch.tensor([0, 0, 0, 0], device='cuda:0')
```

```
gt_instance = detectron2.structures.instances.Instances(image_
size=(800, 800))
gt_instance.set('gt_boxes', boxes)
gt_instance.set('gt_classes', classes)
gt_instances = [gt_instance]
```

Specifically, four objects are in this input image and are all human (class value of 0). After having the ground-truth bounding boxes, the following code snippet performs the sampling strategy described previously:

```
gt_labels, gt_boxes = rpn.label_and_sample_anchors(anchors,
gt_instances)
print(gt_labels[0].shape)
print(gt_boxes[0].shape)
```

This code snippet should give the following output:

```
torch.Size([37500])
torch.Size([37500, 4])
```

The variable gt_labels contain labels (positive=1, negative=0, and ignored=-1) for all 37,500 anchors. The gt_boxes variable contain the coordinates of the ground-truth boxes for the anchors. All these anchors are mapped into one of the four ground-truth boxes in the input. The following code snippet proves this point:

```
np_gt_boxes = gt_boxes[0].to("cpu").numpy()
print(np.unique(np_gt_boxes, axis=0))
```

Specifically, this snippet converts gt_boxes into NumPy and gets their unique bounding boxes in this list. This code snippet should have the following output:

```
[[190. 206. 352. 744.]
 [330. 285. 423. 690.]
 [375. 444. 479. 747.]
 [419. 375. 633. 711.]]
```

These are exactly the bounding boxes of the ground-truth boxes specified previously. Similarly, the following code snippet displays the labels for the sampled anchors:

```
na = gt_labels[0].to("cpu").numpy()
print('positive', sum([x == 1 for x in na]))
```

```
print('negative', sum([x == 0 for x in na]))
print('ignore', sum([x == -1 for x in na]))
```

Specifically, this code snippet gets the labels and counts the positive, negative, and ignored anchors based on labels. This code snippet should display the following output:

```
positive 42
negative 214
ignore 37244
```

Thus, out of 37,500 anchors, 42 are positive, and 214 are negative. These 256 anchors contribute to the loss calculation (training), and the rest are ignored. Now, the RPN loss calculation is the weighted contributions between the classification loss (based on negative and positive labels of the sampled anchors and their respective predictions) and the regression loss (based on the ground-truth bounding box locations and their respective predicted anchor deltas). After all the previous steps, the next task of the RPN is to make the proposals. The following sections inspect the way RPN proposes regions.

Proposal predictions

The proposal prediction step has a new concept called **Non-Maximum Suppression** (**NMS**). Specifically, there can be multiple proposals made for one ground-truth box. These proposals may overlap in a large area and have different objectness scores. *Figure 4.10* shows an example of why NMS is necessary. Specifically, two proposals (dashed boxes) are made for one ground-truth box (solid one). Observably, these two proposals have a large overlapping area, and it is a good idea to remove one and keep another one with a higher objectness score (higher confidence).

Figure 4.10: A ground-truth label (solid box) and multiple proposals (dashed boxes)

The NMS process has the following steps:

1. Select the `pre_nms_topk` proposals with the highest objectness scores. The reason is that the NMS process compares all the proposals pairwise. Thus, it is expensive to perform this step with all the proposals. Instead, it is better to remove proposals with low objectness scores first.

2. Apply IoU computations for every selected proposal pair after *step 1*. If a pair has an IoU greater than a threshold (`nms_thresh`), remove the one with a lower objectness score.

3. After removing overlapped proposals, keep only the `post_nms_topk` number of proposals and forward them to the next step.

The following code snippet shows the configuration parameters for NMS:

```
print('pre_nms_topk train', cfg.MODEL.RPN.PRE_NMS_TOPK_TRAIN)
print('pre_nms_topk test', cfg.MODEL.RPN.PRE_NMS_TOPK_TEST)
print('post_nms_topk train', cfg.MODEL.RPN.POST_NMS_TOPK_TRAIN)
print('post_nms_topk test', cfg.MODEL.RPN.POST_NMS_TOPK_TEST)
print('nms_thresh', cfg.MODEL.RPN.NMS_THRESH)
```

The previous code snippet should give the following output:

```
pre_nms_topk train 12000
pre_nms_topk test 6000
post_nms_topk train 2000
post_nms_topk test 1000
nms_thresh 0.7
```

Specifically, out of 37,500 proposals, it selects the top 12,000 (if training) and 6,000 (if testing) proposals with the highest objectness scores. IoU scores of these selected proposals are then computed and any pair with an IoU score greater than 0.7, and then the one with a lower objectness score is removed. If the remaining proposals are greater than 2,000 (if training) or 1,000 (if testing), order the proposals by objectness scores descendingly and remove the proposals that exceed these numbers. The following code snippet performs the proposal predictions and NMS process as described:

```
pred_objectness_logits = [
            # (N, A, Hi, Wi) -> (N, Hi*Wi*A)
            score.permute(0, 2, 3, 1).flatten(1)
            for score in pred_objectness_logits
        ]
pred_anchor_deltas = [
    # (N, A*B, Hi, Wi) -> (N, Hi*Wi*A, B)
```

```
    x.view(x.shape[0], -1, 4, x.shape[-2], x.shape[-1])
    .permute(0, 3, 4, 1, 2)
    .flatten(1, -2)
    for x in pred_anchor_deltas
]
proposals = rpn.predict_proposals(anchors, pred_objectness_
logits, pred_anchor_deltas, [(800, 800)])
print(proposals[0].proposal_boxes.tensor.shape)
print(proposals[0].objectness_logits.shape)
```

Specifically, this snippet first converts `pred_objectness_logits` and `pred_anchor_deltas` to the appropriate formats before calling the `predict_proposals` function in the RPN head to perform the proposal prediction in the previously described steps. The output of this code snippet should be as follows:

```
torch.Size([1000, 4])
torch.Size([1000])
```

Specifically, there are 1,000 non-overlapping (their IoU scores are smaller than `nms_thresh`) remaining proposals and their corresponding objectness scores. These remaining proposals are then passed into the second stage, *Region of Interest Heads*, to perform the final classification and localization stage. The following section covers details about this Region of Interest Heads.

Region of Interest Heads

The components of the Region of Interest Heads perform the second stage in the object detection architecture that Detectron2 implements. *Figure 4.10* illustrates the steps inside this stage.

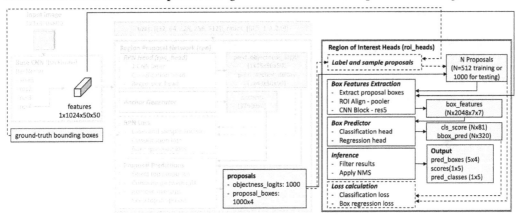

Figure 4.11: The Region of Interest Heads

Specifically, this stage takes the features extracted from the backbone network and the ground-truth bounding boxes (if training) and performs the following steps:

1. Label and sample proposals (if training).

2. Extract box features.

3. Perform predictions.

4. Calculate losses (if training).

5. Perform inferences (if inferencing).

If it is training, out of the 2,000 proposals (`POST_NMS_TOPK_TRAIN`), there can be many negative proposals compared to those positive ones (especially at the early stage of the training when the RPN is not accurate yet). Similar to the RPN stage, this step also labels (based on ground truth) and samples another mini-batch with a fraction of positive proposals for training. Only these sampled proposals contribute to the loss calculation. The following code snippet shows these configuration parameters currently set by Detectron2, and you can also change these parameters if needed:

```
print(cfg.MODEL.ROI_HEADS.IOU_THRESHOLDS)
print(cfg.MODEL.ROI_HEADS.BATCH_SIZE_PER_IMAGE)
print(cfg.MODEL.ROI_HEADS.POSITIVE_FRACTION)
```

This code snippet should display three values `0.5`, `512`, and `0.25`. These values mean that a proposal with IoUs with all ground-truth boxes less than 0.5 is considered background. Per image, 512 proposals are sampled for training, of which 25% should be positive proposals. If there are fewer positive proposals than these, other negative proposals are added to make a total of 512 proposals.

The following code snippet displays the main components for the Region of Interest Head (`roi_head`):

```
roi_heads = predictor.model.roi_heads
print(roi_heads)
```

The output of this code snippet is long. Thus, it is not displayed here. Generally, there are three main components in this stage: `pooler`, `res5`, and `box_predictor`. The `pooler` and `res5` components help to extract box features before completing the prediction with `box_predictor`; the following sections investigate these components.

The pooler

Proposals have different sizes, and the neural networks in the next steps require input features of a common size. Therefore, there must be a way to crop the extracted features into the same size before feeding to the prediction step. The `pooler` component helps perform this task. The following code snippet displays further information about the pooler:

```
print(roi_head.pooler)
```

This code snippet should display the following output:

```
ROIPooler(
  (level_poolers): ModuleList(
    (0): ROIAlign(output_size=(14, 14), spatial_scale=0.0625,
sampling_ratio=0, aligned=True)
  )
)
```

This pooler is an instance of the `ROIPooler` class, which implements the `ROIAlign` (Region of Interest Version 2). Generally, the proposals are specified in the original input image coordinates (800×800 in this case). These coordinates are first translated into the extracted features' size (width×height = 50×50 in this case). Therefore, the spatial scale is 0.0625 (`spatial_scale = 50/800`).

Additionally, the output size for all proposals is specified as 14×14. In ROIAlignV2, after translating the proposal area into the corresponding area in the feature map, this mapped area is divided by a grid of size 14 by 14. Each grid cell is then interpolated (e.g., using the bilinear interpolation method) by four points. The `aligned=True` option means the pixel coordinate is shifted by 0.5 pixels. Half a pixel is not significant in the original image. Still, in the small feature map resolution, it is substantial and helps to improve accuracy (especially in the case of instance segmentation, which we will learn about later). The following code snippet generates the pooler data:

```
x = roi_heads.pooler([res4], [proposals[0].proposal_boxes])
print(x.shape)
```

This code snippet should display a tensor of size (`1000, 1024, 14, 14`) for 1,000 proposals; each has 1,024 channels of 14 by 14 resolution.

After having features for all the proposals of the same size, these proposals undergo another block of convolution layers (called `res5`) to extract features at further depths. The following code snippet performs this salient feature extraction:

```
box_features = roi_heads.res5(x)
print(box_features.shape)
```

This code snippet should display a tensor of size (1000, 2048, 7, 7). Notably, after this layer, features for all the proposals have a smaller resolution of 7×7 and have more salient features (2,048 channels instead of 1,024 channels). The output is called the box features (box_features). These features of the same size are ready for the box predictor described in the following section.

The box predictor

The box predictor takes the box features from the pooler with the same size and performs predictions for box locations and class scores. Executing this print(box_predictor) statement in a code cell displays its layers:

```
FastRCNNOutputLayers(
    (cls_score): Linear(in_features=2048, out_features=81,
bias=True)
    (bbox_pred): Linear(in_features=2048, out_features=320,
bias=True)
)
```

Observably, this box_predictor has two linear (fully connected) layers, one for predicting the classes (cls_score) and another one for predicting the bounding boxes (bbox_pred) for each of the input proposals. The following code snippet performs these predictions:

```
predictions = roi_heads.box_predictor(box_features.mean(dim=(2,
3)))
cls_score, bbox_pred = predictions
print(cls_score.shape)
print(bbox_pred.shape)
```

First, the box features are averaged (mean) for each channel (averaged over the height and width dimensions or dim = (2, 3)) before passing to the two linear layers in the box_predictor. This code snippet should display two tensors of sizes [1000, 81] and [1000, 320]. The first one means 81 class scores per proposal (80 class labels and 1 for the background). The second one means one bounding box (4 location values) per class label (80 classes) per proposal.

If it is training, these class scores and box predictions (of the sampled proposals) are passed to the classification loss calculation and the box regression loss calculation. If it is inferencing, there is another step to perform further processing. Specifically, many predictions have low confidence. Furthermore, these values are currently specified in logits (to avoid issues with numerical approximations for small values) and deltas of the proposals (differences between the predicted boxes and the proposals). Thus, a further processing step is required:

```
pred_instances, _ = roi_heads.box_predictor.
inference(predictions, proposals)
print(pred_instances)
```

Generally, this inferencing step performs the following:

1. Filter results with prediction scores greater than or equal to a threshold (SCORE_THRESH_TEST).

2. Convert the predicted proposal box deltas to box locations in the original images using the bounding boxes of the proposals.

3. Apply NMS for each class independently using a threshold (NMS_THRESH_TEST).

Steps 2 and *3* use two other parameters: the confidence threshold (minimum class score to keep the prediction) and the non-maximum suppression threshold to remove all bounding boxes that have an IoU greater than this threshold and only keep one with the highest class score (confidence). The following code snippet displays these two values:

```
print(cfg.MODEL.ROI_HEADS.SCORE_THRESH_TEST)
print(cfg.MODEL.ROI_HEADS.NMS_THRESH_TEST)
```

This code snippet should display 0.95 and 0.5. The first value dictates that any prediction with a class score of less than 0.95 is removed. The second one implies that for each class, if bounding boxes have an IoU greater than 0.5, all these bounding boxes are removed, and only one with the highest class score (the highest confidence) is kept. These two parameters are configurable.

Finally, after having the output, the following code snippet visualizes the result:

```
from detectron2.utils.visualizer import Visualizer
from detectron2.data import MetadataCatalog
metadata = MetadataCatalog.get(cfg.DATASETS.TRAIN[0])
v = Visualizer(img[:, :, ::-1], metadata)
instances =  pred_instances[0].to("cpu")
annotated_img = v.draw_instance_predictions(instances)
imshow(annotated_img.get_image())
```

This visualization code snippet is similar to what is covered in *Chapter 2*. *Figure 4.11* shows the output of this code snippet.

Figure 4.12: Visualizations of the object detection process

Congratulations! At this point, you should have mastered the typical architecture of Detectron2's implementation of the Faster R-CNN model for the object detection task. This chapter is a little involved. However, the design and philosophies are helpful for you to understand other computer vision tasks and even other popular computer vision techniques such as YOLO.

Summary

This chapter dives deep into the components of Detectron2's implementation of Faster R-CNN for object detection tasks. This model is a two-stage technique: region proposal stage and region of interest extraction stage. Both of these stages use the features extracted from a backbone network. This backbone network can be any state-of-the-art convolutional neural network to extract salient features from the input images. The extracted features and information to generate a set of initial anchors (sizes and ratios; *Chapter 7* explains more about how to customize these sizes and ratios) are then passed to the region proposal neural network to predict a fixed number of proposals with objectness scores (if there is an object in a proposal) and location deltas (location differences between the predicted proposals and the raw anchors). The selected proposals are then passed to the second stage with the region of interest heads to predict the final object classification and localization. Understanding these components in depth prepares you well for model customizations and optimizations to build better models on custom datasets in the next chapters.

Congratulations! By this point, you should have profound knowledge about the architecture of Detectron2 in general and its implementation of Faster R-CNN for object detection models. In the next chapter, you will train custom object detection models using Detectron2.

5

Training Custom Object Detection Models

This chapter starts with an introduction to the dataset and dataset preprocessing steps. It then continues with steps to train an object detection model using the default trainer and an option for saving and resuming training. This chapter then describes the steps to select the best from a set of trained models. It also provides the steps to perform object detection tasks on images with discussions on establishing appropriate inferencing thresholds and classification confidences. Additionally, it details the development process of a custom trainer by extending the default trainer and incorporating a hook into the training process.

By the end of this chapter, you will be able to train Detectron2's models using the default trainer provided by Detectron2 and develop custom trainers to incorporate more customizations into the training process. Specifically, this chapter covers the following topics:

- Processing data
- Using the default trainer
- Selecting the best model
- Developing a custom trainer
- Utilizing the hook system

Technical requirements

You should have completed *Chapter 1* to have an appropriate development environment for Detectron2. All the code, datasets, and results are available in the GitHub repository of the book (under the folder named `Chapter05`) at `https://github.com/PacktPublishing/Hands-On-Computer-Vision-with-Detectron2`. It is highly recommended to download the code and follow along.

> **Important note**
>
> This chapter executes code on a Google Colab instance. You should either map a Google Drive folder and store the output in a mapped folder or download the output and save it for future use. Alternatively, connect these Google Colab notebooks to a local instance, if you have one, to keep the results in permanent storage and utilize better computation resources.

Processing data

The following sections describe the dataset used in this chapter and discuss the typical steps for training Detectron2 models on custom datasets. The steps include exploring the dataset, converting the dataset into COCO format, registering the dataset with Detectron2, and finally, displaying some example images and the corresponding brain tumor labels.

The dataset

The dataset used is the brain tumor object detection dataset available from Kaggle (https://www.kaggle.com/datasets/davidbroberts/brain-tumor-object-detection-datasets), which is downloaded to the GitHub repository of this book to assure its accessibility. This dataset is chosen because medical image processing is a critical subfield in computer vision. At the same time, the task is challenging, and the number of images is appropriate for demonstration purposes.

Downloading and performing initial explorations

The first step in data processing is downloading and performing initial data explorations. The following code snippet downloads this dataset and unzips it into the current working directory:

```
!wget https://github.com/PacktPublishing/Hands-On-Computer-
Vision-with-Detectron2/blob/main/datasets/braintumors.
zip?raw=true -O braintumors.zip
!unzip braintumors.zip -d braintumors
```

The following code snippet then shows us the folders inside this dataset:

```
import os
data_folder = "braintumors"
data_folder_yolo = data_folder + "_yolo"
data_folder_coco = data_folder + "_coco"
folders = os.listdir("braintumors")
print(folders)
```

Specifically, this dataset has three subfolders: `['coronal_t1wce_2_class', 'sagittal_t1wce_2_class', 'axial_t1wce_2_class']`. The next logical step to explore this dataset is to print out its structure using the `tree` command. Therefore, the following code snippets install `tree` and display the folder structure of this dataset:

```
!sudo apt-get install tree
!tree braintumors/ -d
```

Note that the `-d` option is to show the directories only to avoid a long display due to the large numbers of files in these folders. Specifically, the preceding code snippet should display the following folder structure:

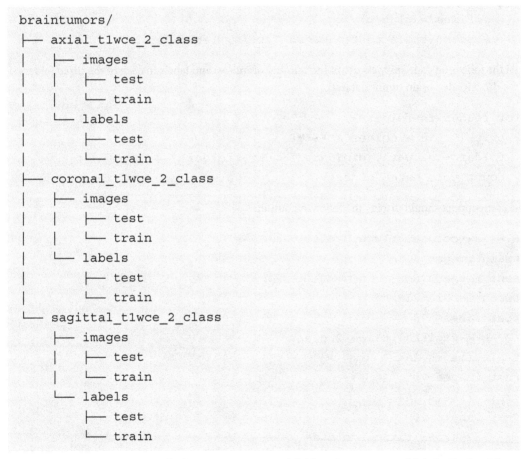

```
braintumors/
├── axial_t1wce_2_class
│   ├── images
│   │   ├── test
│   │   └── train
│   └── labels
│       ├── test
│       └── train
├── coronal_t1wce_2_class
│   ├── images
│   │   ├── test
│   │   └── train
│   └── labels
│       ├── test
│       └── train
└── sagittal_t1wce_2_class
    ├── images
    │   ├── test
    │   └── train
    └── labels
        ├── test
        └── train
```

Observably, each of the three folders contains two subfolders for images and labels. Each of these subfolders has two other folders for the `train` and `test` data. Notably, this dataset is in the YOLO annotation format.

The following code snippet prepares a helper to count the number of images and labels from a dataset with the YOLO annotation format:

```
from glob import glob
def count_yolo_data(folder):
  for images_labels in ["images", "labels"]:
    for train_test in ["train", "test"]:
      file_ext = "*.jpg" if images_labels == "images" else
"*.txt"
        p = os.path.join(folder, images_labels, train_test, file_
ext)
        files = glob(p)
        print(train_test, images_labels, len(files))
```

Then the following code snippets count the number of images and labels in each of the three folders in the downloaded brain tumor dataset:

```
for folder in folders:
  print("-"*8 + folder + "-"*8)
  folder = os.path.join(data_folder, folder)
  count_yolo_data(folder)
```

This code snippet should display the following output:

```
--------coronal_t1wce_2_class--------
train images 319
test images 78
train labels 318
test labels 78
--------sagittal_t1wce_2_class--------
train images 264
test images 70
train labels 264
test labels 70
--------axial_t1wce_2_class--------
train images 310
test images 75
train labels 296
test labels 75
```

Notably, this dataset has a relatively small number of images, and a few do not have labels (there aren't any brain tumors in the pictures).

One approach is to train one model per image type (coronal, sagittal, or axial). However, due to the small number of images, it might be more reasonable to combine images from these folders into one folder. Therefore, the following code snippet prepares a helper method to copy YOLO files from one folder to another:

```
import os
import shutil
from tqdm import tqdm
# move all into one folder
def copy_yolo_files(from_folder, to_folder, images_labels,
train_test):
  from_path = os.path.join(from_folder, images_labels, train_
test)
  to_path = os.path.join(to_folder, images_labels, train_test)
  os.makedirs(to_path, exist_ok = True)
  # get files
  file_ext = "*.jpg" if images_labels == "images" else "*.txt"
  files = glob(os.path.join(from_path,file_ext))
  # move files
  for file in tqdm(files):
    shutil.copy(file, to_path)
```

Then, the following code snippet combines images from three folders into one:

```
# copy
for from_folder in folders:
  from_folder = os.path.join(data_folder, from_folder)
  to_folder = data_folder_yolo
  for images_labels in ["images", "labels"]:
    for train_test in ["train", "test"]:
      copy_yolo_files(from_folder, to_folder, images_labels,
train_test)
```

The following code snippet displays the structure of the newly combined dataset:

```
# Now the folder should look like
!tree -d {data_folder_yolo}
```

This code snippet should have the following output:

```
braintumors_yolo
├── images
│    ├── test
│    └── train
└── labels
     ├── test
     └── train
```

Next, it is reasonable to do some counting of the copied images using the following code snippet:

```
count_yolo_data(data_folder_yolo)
```

This code snippet should display the following output:

```
train images 893
test images 223
train labels 878
test labels 223
```

Notably, this dataset has train and test folders. In several cases, you may want to divide the train folder into train and evaluation (or dev) folders. The evaluation folder selects the best model trained on the training dataset, and the test folder ensures that the model chosen generalizes well in reality. However, this case considers the test dataset as the evaluation set to select the best model from training for demonstration purposes. In other words, this case assumes that the test set represents the future/unseen data well. Therefore, the dataset is kept in the current two folders (train/test sets).

The current YOLO datasets store the metadata (class labels and so on) in a .yaml file. It is simpler to extract this and keep the classes (or labels) in classes.txt for the newly created YOLO dataset using the following code snippet:

```
import yaml
from yaml.loader import SafeLoader
with open(os.path.join(data_folder, folders[0], folders[0]+'.
yaml')) as f:
   classes = yaml.load(f, Loader=SafeLoader)["names"]
with open(os.path.join(data_folder_yolo, "classes.txt"), "w")
as f:
   f.write('\n'.join(classes))
```

Now we are ready to convert these datasets into COCO format using the code snippet covered in *Chapter 3*.

Data format conversion

Detectron2 supports COCO data format, and the current datasets are in YOLO format. Therefore, the following code snippets perform the data format conversion. First, run the following code snippet to install the `pylabel` package:

```
!pip install -q pylabel
```

Next, the following snippet creates a helper method to perform data conversion (as covered in *Chapter 3*):

```
from pylabel import importer
def yolo_to_coco(input_folder, output_folder, train_test):
  # Some codes are removed here (please refer to GitHub)
  # copy annotations
  for f in tqdm(txt_files):
    shutil.copy(f, coco_dir)
  # copy images
  for f in tqdm(img_files):
    shutil.copy(f, coco_dir)
  # get the classes
  with open(os.path.join(input_folder, "classes.txt"), "r") as
f:
    classes = f.read().split("\n")
  # import
  dataset = importer.ImportYoloV5(path=coco_dir, cat_
names=classes, name="brain tumors")
  # export
  coco_file = os.path.join(coco_dir, "_annotations.coco.json")
  # Detectron requires starting index from 1
  dataset.export.ExportToCoco(coco_file, cat_id_index=1)
  # now delete yolo annotations in coco set
  for f in txt_files:
    os.remove(f.replace(labels_path, coco_dir))
```

Specifically, this code snippet copies images and annotations to the `coco_dir` folder, imports the YOLO dataset, exports it to the COCO dataset, and removes the YOLO annotation files. Next, the following code snippet performs the actual data conversion and copies the resulting datasets into a new folder called `braintumors_coco`:

```
yolo_to_coco(data_folder_yolo, data_folder_coco, "train")
yolo_to_coco(data_folder_yolo, data_folder_coco, "test")
```

Running the `!tree -d {data_folder_coco}` command in a code cell should display the structure of the resulting datasets as the following:

```
braintumors_coco
├── test
└── train
```

Now the dataset is in COCO format. It is reasonable to test and display some images and their corresponding labels.

Displaying samples

There are several ways to display data samples. However, it is more convenient to use Detectron2's visualization utility to perform this task. Thus, we first need to install Detectron2 using the following code snippet:

```
!python -m pip install \
'git+https://github.com/facebookresearch/detectron2.git'
```

Next, we import the `register_coco_instances` method and utilize it to register the train/test datasets:

```
from detectron2.data.datasets import register_coco_instances
# Some simple variable settings are removed
# Register datasets
register_coco_instances(
    name = name_ds_train,
    metadata = {},
    json_file = json_file_train,
    image_root = image_root_train)
register_coco_instances(
    name = name_ds_test,
    metadata = {},
```

```
    json_file = json_file_test,
    image_root = image_root_test)
```

The following code snippet displays three sample images and their corresponding labels from the `train` dataset:

```
plot_random_samples(name_ds_test, n = 3)
```

The output of this code snippet is shown in *Figure 5.1*:

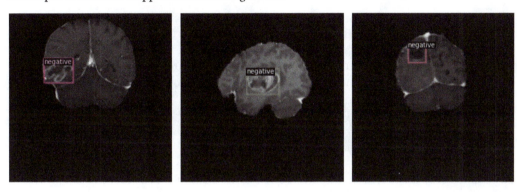

Figure 5.1: Sample images and corresponding labels from the train dataset

Similarly, the following code snippet displays another three sample images and their corresponding labels from the `test` dataset:

```
plot_random_samples(name_ds_test, n = 3)
```

The resulting output of this code snippet is shown in *Figure 5.2*:

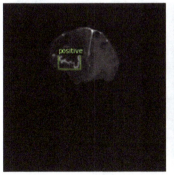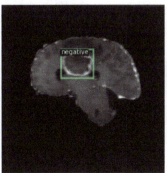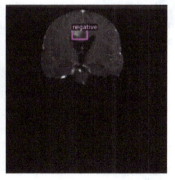

Figure 5.2: Sample images and corresponding labels from the test dataset

At this point, if all these data processing steps are performed on Google Colab, it is a good time to download the resulting COCO datasets to your local computer or to map the Google Drive folder and save the resulting datasets into a Google Drive folder to avoid losing the datasets after the Google Colab instance is turned off. In this specific case, the converted COCO datasets are stored in the GitHub folder of this book for future sections.

Now that the datasets are ready and correctly registered, it is time to train models and see how they perform using Detectron2. The simplest way to perform model training on custom datasets is to use the default trainer class provided by Detectron2.

Using the default trainer

Detectron2 provides a default trainer class, which helps to train Detectron2 models on custom datasets conveniently. First, we download the datasets converted in the previous section and unzip them:

```
!wget -q https://github.com/PacktPublishing/Hands-On-Computer-
Vision-with-Detectron2/raw/main/datasets/braintumors_coco.zip
!unzip -q braintumors_coco.zip
```

Next, install Detectron2 and register the train/test datasets using the exact code snippets provided in the previous section. Additionally, before training, run the following code snippet to prepare a logger that Detectron2 uses to log training/inferencing information:

```
from detectron2.utils.logger import setup_logger
logger = setup_logger()
```

After having the datasets registered and setting up the logger, the next step is getting a training configuration. Precisely, we set the output directory (where we will store the logging events and the trained models), the path to store the configuration file to be used in the future inferences, the number of classes as positive and negative (nc = 2), and the device option as cuda (for training on GPUs). Finally, to speed up training, we also select the ResNet50 FPN as the backbone network for the model to be trained:

```
import os
from detectron2.config import get_cfg
from detectron2 import model_zoo
import pickle

output_dir = "output/object_detector"
os.makedirs(output_dir, exist_ok=True)
output_cfg_path = os.path.join(output_dir, "cfg.pickle")
nc = 2
```

```
device = "cuda"

config_file_url = "COCO-Detection/faster_rcnn_R_50_FPN_3x.yaml"
checkpoint_url = "COCO-Detection/faster_rcnn_R_50_FPN_3x.yaml"
```

Additionally, the following code snippet creates a training configuration:

```
# Create a configuration file
cfg = get_cfg()
config_file = model_zoo.get_config_file(config_file_url)
cfg.merge_from_file(config_file)
cfg.MODEL.WEIGHTS = model_zoo.get_checkpoint_url(checkpoint_
url)
# Download weights
cfg.MODEL.WEIGHTS = model_zoo.get_checkpoint_url(checkpoint_
url)
# Set datasets
cfg.DATASETS.TRAIN = (name_ds_train,)
cfg.DATASETS.TEST = (name_ds_test,)
# Workers
cfg.DATALOADER.NUM_WORKERS = 2
# Images per batch
cfg.SOLVER.IMS_PER_BATCH = 8
# Learning rate
cfg.SOLVER.BASE_LR = 0.00025
# Iterations
cfg.SOLVER.MAX_ITER = 5000
cfg.SOLVER.CHECKPOINT_PERIOD = 500
# Classes
cfg.MODEL.ROI_HEADS.NUM_CLASSES = nc
cfg.MODEL.DEVICE = device
cfg.OUTPUT_DIR = output_dir
```

We set the train and test datasets using the corresponding names registered with Detectron2. Next, we set the number of workers for the data loader as 2 and the number of images per batch as 8 (you can change this number to fit the available GPUs). Additionally, since we are using transfer learning (by reusing the weights trained from the Detectron2 Model Zoo), we set a relatively small learning rate – in this case, 0.00025. Furthermore, we will train for 5,000 iterations and save a resulting model every 500 iterations.

Future executions (such as loading models for inferences) need the configuration file to load trained model weights. Therefore, it is important to save this file. The following code snippet saves the generated configuration file for future uses:

```
# save configuration file for future use
with open(output_cfg_path, "wb") as f:
    pickle.dump(cfg, f, protocol = pickle.HIGHEST_PROTOCOL)
```

Finally, it is relatively easy to use the `DefaultTrainer` class from Detectron2 to start training a custom model using the configuration file:

```
from detectron2.engine import import DefaultTrainer
trainer = DefaultTrainer(cfg)
trainer.train()
```

Notably, Google Colab may stop its execution for several reasons (e.g., out of execution quota or idle time-out). Therefore, Detectron2 provides an option to resume training by adding a `trainer.resume_or_load(True)` method before calling to `trainer.train()` when you continue training from the previous one. This statement dictates that the trainer load the latest weights from the previous run, if there are any, and continue from there.

The training process may take from one to a few hours, depending on the available computation resources. You can view the **eta** (**estimated time to arrival**) to know how long it will take to complete the training. Once the training finishes, the output models and the log events are stored in the configured output folder (`cfg.OUTPUT_DIR`). Specifically, running the `!ls -1 {cfg.OUTPUT_DIR}` statement in a code cell brings up the following output:

```
cfg.pickle
events.out.tfevents.1671405675.phamserver.266249.0
last_checkpoint
metrics.json
model_0000499.pth
model_0000999.pth
model_0001499.pth
model_0001999.pth
model_0002499.pth
model_0002999.pth
model_0003499.pth
model_0003999.pth
model_0004499.pth
```

```
model_0004999.pth
model_final.pth
```

The output includes the configuration file (`cfg.pickle`), a file for log events, the last checkpoint (`last_checkpoint`) so the model can resume its training if needed, a file for all the training evaluation metrics (`metrics.json`), and all the models at every 500 (`cfg.SOLVER.CHECKPOINT_PERIOD`) iterations. Notably, the model at the last iteration (`cfg.SOLVER.MAX_ITERATIONS`) is stored as `model_final.pth`.

After having the training results, the next step is to perform some model evaluations and select the best-performing model.

Selecting the best model

Selecting the best model requires evaluation metrics. Therefore, we need to understand the common evaluation terminologies and evaluation metrics used for object detection tasks before choosing the best model. Additionally, after having the best model, this section also covers code to sample and visualize a few prediction results to qualitatively evaluate the chosen model.

Evaluation metrics for object detection models

Two main evaluation metrics are used for the object detection task: mAP@0.5 (or AP50) and F1-score (or F1). The former is the mean of average precisions (mAP) at the intersection over the union (IoU) threshold of 0.5 and is used to select the best models. The latter represents the harmonic means of precision and recall and is used to report how the chosen model performs on a specific dataset. The definitions of these two metrics use the computation of *Precision* and *Recall*:

$$Precision = \frac{TP}{TP + FP}$$

$$Recall = \frac{TP}{TP + FN}$$

Here, **TP** (for **True Positive**) means the number of predicted objects that are correctly classified and localized (based on the IoU threshold), **FP** (for **False Positive**) represents the number of predicted objects that are incorrectly classified or localized, and **FN** (for **False Negative**) means the number of ground-truth objects that are not correctly predicted. F1-score can then be computed using the following formula:

$$F1 = \frac{2 \times Precision \times Recall}{Precision + Recall}$$

F1-score balances the precision and recall and measures how a model performs on a specific dataset. However, to select a general model that generalizes well on unseen data, mAP@0.5 is more often used. The definition of mAP@0.5 is a little involved. *Figure 5.3* illustrates this definition by dividing it into parts:

Figure 5.3: Intersection over Union (IoU) and mean Average Precision at IoU=0.5 (mAP@0.5)

Specifically, the IoU is computed as the ratio of the overlapping region between the predicted bounding box and the ground-truth bounding box. These two boxes are considered the same if the IoU value is greater than or equal to 0.5. Next, for each class label (category), we compute the **average precision** (**AP**) values for different recall values (typically ranging from 0 to 1 with a step value of 0.1). Finally, the **mean AP** (**mAP**) value across different class categories is the mAP@0.5 evaluation metric. Therefore, this measurement is robust and helps to select a model that should scale well to the unseen datasets.

Selecting the best model

In this case, we do not evaluate how our model performs on a specific test set but select the best model that scales well for the unseen datasets. Therefore, mAP@0.5 is used. The following code snippet selects the trained models:

```
model_names = []
max_iter = cfg.SOLVER.MAX_ITER
chp = cfg.SOLVER.CHECKPOINT_PERIOD
for i in range(1, max_iter//chp):
  model_names.append(f'model_{str(i*chp-1).zfill(7)}.pth')
model_names.append('model_final.pth')
print(model_names)
```

The output of this code snippet should look like the following:

```
['model_0000499.pth',
 'model_0000999.pth',
 'model_0001499.pth',
 'model_0001999.pth',
 'model_0002499.pth',
```

```
  'model_0002999.pth',
  'model_0003499.pth',
  'model_0003999.pth',
  'model_0004499.pth',
  'model_final.pth']
```

This code snippet lists all the models saved at every checkpoint period (cfg.SOLVER.CHECKPOINT_PERIOD) over the training process. Notably, the model after the training completes (at the cfg.SOLVER.MAX_ITER iteration) is named model_final.pth. Additionally, the number in the model name is always padded with zeros to have a length of 7.

The following code snippet prepares an evaluator to evaluate how each of these models performs on the test set:

```
from detectron2.engine import DefaultPredictor
from detectron2.evaluation import COCOEvaluator
evaluator = COCOEvaluator(
    dataset_name = name_ds_test,
    tasks=("bbox",),
    distributed=False,
    output_dir = os.path.join(output_dir, 'test_results')
    )
```

Notably, COCOEvaluator is selected because the dataset is in COCO format. Additionally, the "bbox" input for the tasks option indicates to the evaluator to perform evaluations for object detection tasks. The test results are stored in the 'test_results' folder under the output directory. After getting the evaluator, the following code snippet performs the evaluations on each of the generated models:

```
# evaluate all folders
import logging
logger.setLevel(logging.CRITICAL)
model_eval_results = []
for model_name in model_names:
  # load weights
  cfg.MODEL.WEIGHTS = os.path.join(cfg.OUTPUT_DIR, model_name)
  cfg.MODEL.ROI_HEADS.SCORE_THRESH_TEST = 0.5
  # predictor
  predictor = DefaultPredictor(cfg)
  # evaluate
  x = trainer.test(cfg, predictor.model,
```

```
evaluators=[evaluator])
  model_eval_results.append(x['bbox'])
```

Notably, this snippet first sets the logging level to CRITICAL to reduce the output messages and focus on detection metrics only. This code snippet produces evaluation results for all the models. Typically, one model should have the values for the following evaluation metrics:

```
Average Precision  (AP) @[ IoU=0.50:0.95 | area=    all |
maxDets=100 ] = 0.281
 Average Precision  (AP) @[ IoU=0.50      | area=    all |
maxDets=100 ] = 0.383
 Average Precision  (AP) @[ IoU=0.75      | area=    all |
maxDets=100 ] = 0.340
 Average Precision  (AP) @[ IoU=0.50:0.95 | area= small |
maxDets=100 ] = 0.251
 Average Precision  (AP) @[ IoU=0.50:0.95 | area=medium |
maxDets=100 ] = 0.284
 Average Precision  (AP) @[ IoU=0.50:0.95 | area= large |
maxDets=100 ] = 0.647
 Average Recall     (AR) @[ IoU=0.50:0.95 | area=    all |
maxDets=  1 ] = 0.473
 Average Recall     (AR) @[ IoU=0.50:0.95 | area=    all |
maxDets= 10 ] = 0.508
 Average Recall     (AR) @[ IoU=0.50:0.95 | area=    all |
maxDets=100 ] = 0.508
 Average Recall     (AR) @[ IoU=0.50:0.95 | area= small |
maxDets=100 ] = 0.482
 Average Recall     (AR) @[ IoU=0.50:0.95 | area=medium |
maxDets=100 ] = 0.524
 Average Recall     (AR) @[ IoU=0.50:0.95 | area= large |
maxDets=100 ] = 0.733
```

There are 12 COCO evaluation metrics. However, generally, mAP@0.5 or Average Precision (AP) @[IoU=0.50 | area = all | maxDets=100] is used for selecting the best model (you may change this to meet individual needs). Therefore, the following code snippet extracts these values for all models, gets the best one, and plots these values on a figure for illustration purposes:

```
# some import statements are removed
aps = [x['AP50'] if not math.isnan(x['AP50']) else 0 for x in
model_eval_results]
best_model_idx = np.argmax(aps)
```

```
best_model_name = model_names[best_model_idx]
best_ap = aps[best_model_idx]
anno_text = f'The best model {best_model_name} has
mAP@0.5={round(best_ap, 2)}'

plt.figure(figsize=(12, 6))
x = [i for i in range(len(aps))]
y = aps
plt.xticks(ticks=x, labels=model_names, rotation=45)
plt.plot(x, y)
plt.scatter(x, y)
plt.plot([best_model_idx, best_model_idx], [0, best_ap], '--')
plt.ylabel("mAP@0.5")
plt.text(best_model_idx+0.1, best_ap, anno_text, ha="left",
va="center")
plt.show()
```

Figure 5.4 displays the output of this code snippet:

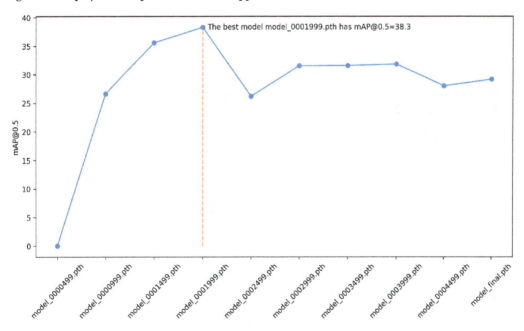

Figure 5.4: The mean average precision (mAP@50) of all the models

Observably, the best model (`model_0001999`, although it might be a different one for you due to having randomness in the training), which is at iteration 2,000, has a mAP@0.5 value of 38.3%. Again, it is important to emphasize that this value is not used to indicate how the model performs on a specific test but to select a relatively better model among a set of models. Furthermore, the models are evaluated against one single confidence score threshold (`SCORE_THRESH_TEST`) of 0.5. However, changing this threshold does not impact the relative mAP@0.5 values among models. The following section illustrates this idea in further detail.

Inferencing thresholds

The following code snippet evaluates the impact of this confidence threshold for the selected model (`model_0001999`):

```
thresh_tests = [0.1*i for i in range(10)]
thresh_eval_results = []
for thresh_test in thresh_tests:
    cfg.MODEL.WEIGHTS = os.path.join(cfg.OUTPUT_DIR, best_model_
name)
    cfg.MODEL.ROI_HEADS.SCORE_THRESH_TEST = thresh_test
    predictor = DefaultPredictor(cfg)
    x = trainer.test(cfg, predictor.model,
evaluators=[evaluator])
    thresh_eval_results.append(x['bbox'])
```

The visualization code snippet similar to the previous section can be used to extract mAP@0.5 for different confidence thresholds and plot a graph for illustration purposes, as shown in *Figure 5.5*:

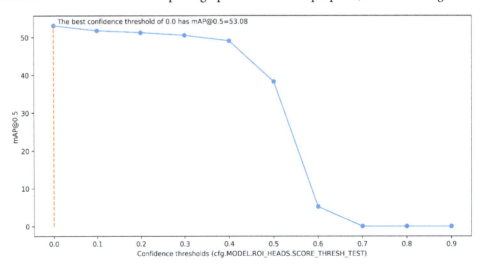

Figure 5.5: The impact of the confidence threshold on mean average precision (mAP@0.5)

Generally, reducing the confidence threshold brings more recalls with fewer precisions and leads to having higher mAP@0.5. However, a smaller threshold also makes the inference slower (due to having many regions of interest before going through the NMS process).

It is worth noting that this behavior is generally true for all models. Therefore, the model selection code snippet fixes the threshold (0.5) previously. Additionally, it does not mean we should reduce this threshold to have a higher mAP@0.5 while assessing how a model works on a specific dataset. Generally, to determine how a model works on a particular dataset, F1-score is used instead. Therefore, setting too low a confidence threshold may bring too many false positives and reduce the F1 score. Consequently, we should perform a similar analysis by changing this score threshold and see which one yields the highest F1 score. For the brief of this section, we skip the F1 score evaluation. However, the analysis remains similar to the code provided here. Once you have the best model, the following section shows some sample predictions on the test dataset.

Sample predictions

The following code snippet prepares a helper method that samples random images from a given dataset, performs the predictions on the sampled images, and displays the images with the ground-truth labels and predicted labels:

```python
# Some import statements are removed for space efficiency
def plot_random_samples(name_ds, n=3, predictor=None):
    ds = DatasetCatalog.get(name_ds)
    met = MetadataCatalog.get(name_ds)
    nrows, ncols = int(-(-n/3)),3
    samples = random.sample(ds, n)
    fs = (21, 17)
    fig,axs= plt.subplots(nrows=nrows,ncols=ncols,figsize=fs)
    for i,s in enumerate(samples):
        ri = i//ncols
        ci = i%ncols
        ax = axs[ri][ci] if len(axs.shape) == 2 else axs[i]
        img = cv2.imread(s["file_name"])
        v = Visualizer(img[:,:, ::-1], metadata=met, scale=0.5)
        # visualize ground-truths
        v = v.draw_dataset_dict(s)
        ax.imshow(v.get_image())
        ax.axis("off")
    plt.tight_layout()
```

```
plt.show()
# Similar code for predicted results (removed here)
```

Notably, the first part of the method samples a few inputs and visualizes the input images and their annotations. The second section (removed) performs predictions on the sampled input and visualizes the corresponding predicted annotations. Next, we create a predictor for the best model using the following code snippet:

```
cfg.MODEL.ROI_HEADS.SCORE_THRESH_TEST  = 0.5
cfg.MODEL.WEIGHTS = os.path.join(cfg.OUTPUT_DIR, best_model_
name)
predictor = DefaultPredictor(cfg)
```

Notably, this code snippet sets the SCORE_THRES_TEST value to 0.5 (which should be optimized using the similar code snippet provided previously) to yield better results. Finally, we can run the following code snippet to randomly sample images and visualize the ground-truth labels and the predicted results:

```
plot_random_samples(name_ds_test, predictor = predictor)
```

This code snippet should display the output, as shown in *Figure 5.6*. Notably, the colors of the bounding boxes are generated by the Detectron2 visualizer, and they do not have meaning in this case.

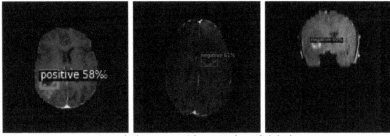

Sampled images with ground-truth labels

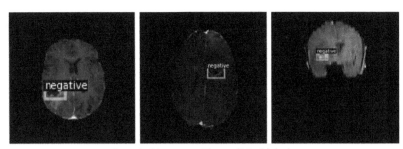

Sampled images with predicted labels

Figure 5.6: Sampled images and their corresponding ground-truth labels and predicted labels

Generally, the localizations (bounding boxes) are reasonably good, while there are still misclassifications. This result is logical because it is easier to observe whether a tumor exists in an image, but it is harder to confirm whether one is positive or negative. Additionally, this model may predict one tumor as positive and negative because the NMS process is currently done per class (as discussed in the previous chapter). However, in this case, a tumor is either positive or negative (not both). Therefore, there is an opportunity to optimize the model by performing the NMS process regardless of the class labels, which we will perform in the next chapter.

Training models using the Detectron2 default trainer (DefaultTrainer) is convenient and can satisfy most common needs. However, in several cases, we may need to develop a custom trainer to facilitate training and analyze the training metrics and results. The following section covers essential components of a trainer class and how to customize them.

Developing a custom trainer

There are several reasons for developing a Detectron2 custom trainer. For instance, we may want to customize the dataset loader to incorporate more image augmentation techniques or to add evaluators to assess how the trained models perform during training. The following code snippet covers the source code to build a custom trainer for the latter, and *Chapter 8* covers the code for the former:

```python
from detectron2.engine import DefaultTrainer
from detectron2.evaluation import COCOEvaluator
class BrainTumorTrainer(DefaultTrainer):
  @classmethod
  def build_evaluator(cls, cfg, dataset_name, output_
folder=None):
    if output_folder == None:
      output_folder = cfg.OUTPUT_DIR
    else:
      output_folder = os.path.join(cfg.OUTPUT_DIR,
                                   output_folder)
    os.makedirs(output_folder)
    return COCOEvaluator(dataset_name,
                         distributed = False,
                         output_dir = output_folder)
```

Specifically, this custom trainer overrides the `build_evaluator` method to return an object of `COCOEvaluator` that can be used to evaluate the `test` set. The `COCOEvaluator` class is chosen because the `test` dataset is in COCO format. When the evaluator is specified, the following code snippet specifies the frequency of how often this evaluator is executed:

```
cfg.TEST.EVAL_PERIOD = cfg.SOLVER.CHECKPOINT_PERIOD
```

This code snippet sets the evaluation period the same as the checkpoint period, so the evaluation is done for every checkpoint. The next code snippet then creates a custom trainer and performs training just like the default trainer:

```
trainer = BrainTumorTrainer(cfg)
trainer.train()
```

You may wonder why executing an evaluator during the training process is more valuable than evaluating the trained models after the training process is completed. There are several reasons for this usage. One is that this approach allows monitoring evaluation results with COCO evaluation metrics that can be used to select the best models while training and perform early stopping (stopping training after several iterations when you no longer see improvements).

Another example is that the evaluation metrics are stored together with the training metrics (such as classification and bounding box losses) and offer the opportunity to evaluate the bias and variance trade-off during training using tools such as TensorBoard. The following section introduces a new concept called the **hook system** combined with the idea proposed in this section to demonstrate the first scenario. *Chapter 6* utilizes TensorBoard to illustrate the second scenario.

Utilizing the hook system

A hook system allows incorporating classes to execute several tasks on training events. A custom hook builds upon inheriting a base class from Detectron2 called `detectron2.engine.HookBase`. A hook allows the developer to execute tasks on four events by overriding the following methods:

- `before_training()` to include tasks to be executed before the first training iteration
- `after_training()` to include tasks to be executed after training completes
- `before_step()` to include tasks to be executed before each training iteration
- `after_step()` to include tasks to be executed after each training iteration

The following code snippet creates a hook to read the evaluation metrics generated by COCOEvaluator from the previously built custom trainer, keeps track of the best model with the highest mAP@0.5 value, and saves the model as model_best.pth:

```python
# Some import statements are removed for space efficiency
class BestModelHook(HookBase):
    def __init__(self, cfg,
                 metric    = "bbox/AP50",
                 min_max = "max"):
        # Explained next
    def _take_latest_metrics(self):
        # Explained next
    def after_step(self):
        # Explained next
```

This class overrides the __init__ method to pass in metric and another parameter (min_max) to indicate how to monitor the metric value (min or max):

```python
self._period    = cfg.TEST.EVAL_PERIOD
self.metric    = metric
self.min_max    = min_max
self.best_value    = float("-inf") if min_max == "max" else
float("inf")
logger = logging.getLogger("detectron2")
logger.setLevel(logging.DEBUG)
logger.propagate    = False
self._logger    = logger
```

This BestModelHook class also creates a method (_take_latest_metrics) to achieve all the evaluation metrics stored during the training process:

```python
with torch.no_grad():
    latest_metrics = self.trainer.storage.latest()
    return latest_metrics
```

The `after_step` method then loops through all the evaluation metrics and gets the monitoring value, and saves the training model if it has a better evaluation result than the current best one:

```
next_iter = self.trainer.iter + 1
is_final = next_iter == self.trainer.max_iter
if is_final or (self._period > 0 and next_iter % self._period
== 0):
  latest_metrics = self._take_latest_metrics()
    for (key, (value, iter)) in latest_metrics.items():
      if key == self.metric:
        if (self.min_max == "min" and value < self.best_value)
or (self.min_max == "max" and value > self.best_value):
          self._logger.info("Updating best model at iteration
{} with {} = {}".format(iter, self.metric, value))
          self.best_value = value
          self.trainer.checkpointer.save("model_best")
```

Notably, the hook is built by overriding the `after_step()` method. However, it only performs its tasks at every `cfg.TEST.EVAL_PERIOD` iteration when there are results from COCOEvaluator.

The following code snippet utilizes the custom trainer and the hook to perform model evaluation and selection during training:

```
trainer = BrainTumorTrainer(cfg)
bm_hook = BestModelHook(cfg, metric="bbox/AP50", min_max="max")
trainer.register_hooks(hooks=[bm_hook])
trainer.train()
```

Observably, it is relatively easy to register the hook into the trainer. Additionally, the `latest_metrics` variable contains many training and evaluation metrics, and we are extracting only the mAP@0.5 (`bbox/AP50`) on the `test` set. Finally, the training remains the same as the default trainer. At the end, besides the models generated by the default trainer, there is a model called `model_best`. `pth` that stores the weights for the model with the highest mAP@0.5 evaluated on the `test` dataset.

Congratulations! By now, you should have mastered the steps to perform data processing and train custom models on a custom dataset. Additionally, you should understand how to incorporate more custom processing into training.

Summary

This chapter discussed the steps to explore, process, and prepare a custom dataset for training object detection models using Detectron2. After processing the dataset, it is relatively easy to register the train, test, and evaluation data (if there is any) with Detectron2 and start training object detection models using the default trainer. The training process may result in many models. Therefore, this chapter provided the standard evaluation metrics and approaches for selecting the best model. The default trainer may meet the most common training requirements. However, in several cases, a custom trainer may be necessary to incorporate more customizations into the training process. This chapter provided code snippets to build a custom trainer that incorporates evaluations on the test set during training. It also provided a code snippet for a custom hook that extracts the evaluation metrics and stores the best model during training.

The next chapter, *Chapter 6*, uses TensorBoard to analyze the training and evaluation metrics generated by the default trainer and the custom trainers built in this chapter. Analyzing these metrics should offer ways to fine-tune object detection models to achieve higher performance.

Inspecting Training Results and Fine-Tuning Detectron2's Solvers

This chapter covers how to use TensorBoard to inspect training histories. It utilizes the code and visualization approach to explain the concepts behind Detectron2's solvers and their hyperparameters. The related concepts include gradient descent, stochastic gradient descent, momentum, and variable learning rate optimizers. This chapter also provides code to help you find the standard hyperparameters for Detectron2's solvers.

By the end of this chapter, you will be able to use TensorBoard to analyze training results and find insights. You will also have a deep understanding of the essential hyperparameters for Detectron2's solvers. Additionally, you will be able to use code to generate appropriate values for these hyperparameters on your custom dataset. Specifically, this chapter covers the following topics:

- Inspecting training histories with TensorBoard
- Understanding Detectron2's solvers
- Fine-tuning the learning rate and batch size

Technical requirements

You must have completed *Chapter 1* to have an appropriate development environment for Detectron2. All the code, datasets, and results are available in this book's GitHub repository at https://github. com/PacktPublishing/Hands-On-Computer-Vision-with-Detectron2. It is highly recommended that you download the code and follow along.

Inspecting training histories with TensorBoard

Before fine-tuning models, it is essential to understand the evaluation metrics that are measured during training, such as training losses (for example, classification losses and localization losses), learning rate changes, and validation measurements on the test set (for example, mAP@0.5), which change over time. These metrics allow us to understand the training processes for debugging and fine-tuning models. Detectron2's logger automatically logs the evaluation metrics for training and testing data in a format that is usable by TensorBoard.

TensorBoard (https://www.tensorflow.org/tensorboard) offers tools and visualizations so that we can explore training and evaluation processes in machine learning experiments. The following code snippet downloads the training logs for the two experiments we carried out in *Chapter 5* and unzips them:

```
# download and extract
!wget {url_to_object_detector.zip}
!wget {url_to_object_detector_hook.zip}
!unzip object_detector.zip
!unzip object_detector_hook.zip
```

Once you have the training and evaluation logs, the following code snippet loads TensorBoard so that you can view the training results:

```
output_dir = "output/"
%load_ext tensorboard
%tensorboard --logdir {output_dir} --host localhost
```

As we can see, this code snippet sets the output directory to the output folder where the training logs are stored. It then loads TensorBoard as an internal web server and embeds it in the output cell of Google Colab.

> **Important note**
> If TensorBoard was loaded, run `%reload_ext tensorboard` instead of `%load_ext tensorboard` or kill the current server using its process ID before loading a new instance.

Figure 6.1 shows the logs in the **SCALARS** tab for the training and evaluation logs for Detectron2's object detection models:

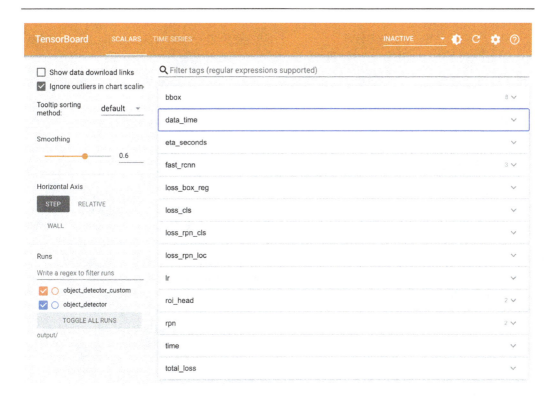

Figure 6.1: TensorBoard and the SCALARS tab for training and evaluation logs

The logging step is done every 20 iterations during training (this number is configurable), and, by default, there are 12 categories of scalars:

- The `data_time` category stores the time it has taken to prepare the data for training.

- The `eta_seconds` category stores the estimated number of seconds until training completes.

- The `fast_rcnn` category stores the Fast R-CNN evaluation metrics. These are classification accuracy (`cls_accuracy`), false negative (`false_negative`), and foreground (object) classification accuracy (`fg_cls_accuracy`).

- The `loss_box_reg` category stores the loss for box regression (or the localization loss).

- The `loss_cls` category stores the classification loss.

- The `loss_rpn_cls` category stores the region proposal network objectness classification loss.

- The `loss_rpn_loc` category stores the region proposal network localization (bounding-box) loss.

- The `lr` category stores the learning rate. Learning rates may vary due to having a warm-up period, steps (reduced at some steps), or using other parameters.

- The `roi_head` category stores the number of background samples (`num_bg_samples`) and the number of foreground samples (`num_fg_samples`).

- The `rpn` category stores the number of negative and positive anchors (`num_neg_anchors` and `num_pos_anchors`, respectively).

- The `time` category stores the clock time (timestamp) and the relative time from the start of the training.

- The `total_loss` category stores the total of all losses.

These are the most common scalars that are evaluated on the training dataset that help us inspect the training process. However, you can write custom code to add log values to the outputs and view/analyze them using TensorBoard. For instance, the custom trainer we developed in *Chapter 5* has a `COCOEvaluator` class that adds COCO evaluation metrics evaluated on the test dataset to the `bbox` category. This category contains eight evaluation metrics (please refer to the previous chapter if you are unfamiliar with the definitions of these evaluation metrics): average precision (`AP`), mAP@0.5 (`AP50`), mAP@0.75 (`AP75`), average precision for the negative cases (`AP-negative`), average precision for the positive cases (`AP-positive`), average precision for the large objects (`API`), average precision for the medium objects (`APm`), and average precision for the small objects (`APs`). *Figure 6.2* shows two of them:

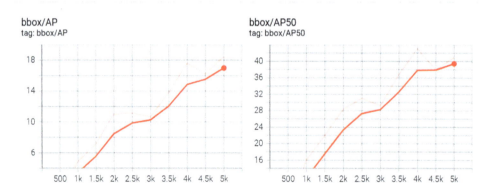

Figure 6.2: Two example COCO evaluation metrics evaluated on the test dataset

Due to the stochastic nature of minibatch training and the small batch size due to our GPU limitation (only four images per batch), the log values might be zigzagging from iteration to iteration (the faded line). Therefore, TensorBoard provides an option to smoothen the log values and see general trends. This smoothing technique helps to show increasing evaluation metrics for all three displayed metrics over time. Specifically, even though AP50 has the highest value at iteration 4,000 (3,999 to be exact, due to the counting starting from 0), the smoothened line indicates that further training iterations may help improve performance.

Currently, mAP@0.5 is used to evaluate the model's performance. However, comparing mAP@0.5 on the test set with the losses on the training set might be challenging while analyzing the bias-variance trade-off. Therefore, writing another custom hook, as in *Chapter 5*, may be helpful to evaluate losses on the test set at every evaluation interval. Detectron2 does not provide this hook by default. Still, it is relatively easy to write one if needed by taking the evaluation code used on the training dataset, as in Detectron2's GitHub repository, running it on the test set, and logging the results for analysis.

Notably, the current TensorBoard instance loads the training outputs from two training experiments from *Chapter 5*: the default trainer and the custom trainer with a COCO evaluator. Both have the evaluation metrics for the other 12 categories, but only the latter has the bbox category due to having the COCO evaluator. On the other hand, the total_loss category has values for both experiments, as shown in *Figure 6.3* (the blue and orange lines):

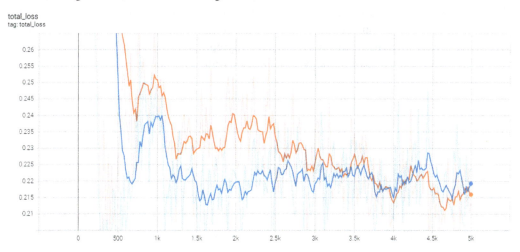

Figure 6.3: Total loss for two different experiments

In both cases, the smoothened lines for training losses indicate the fluctuations of the losses. Therefore, a larger batch size would help if there are available computation resources.

Congratulations! By now, you should be able to quickly and conveniently load the training logs on TensorBoard and analyze the training and evaluation results. Based on the analysis results, we should be able to try some other fine-tuning techniques.

Understanding Detectron2's solvers

We should try two obvious fine-tuning techniques: changing the backbone network and increasing the batch size. As indicated in the Detectron2 Model Zoo (introduced in *Chapter 3*), the backbone we selected in *Chapter 5* is the simplest one (ResNet50), with a low mAP@0.5 on the pre-trained dataset. It is lightweight and fast to train and infer, so we have been using it for our experiments using a free Google Colab plan. If computation resources are available, selecting a more powerful backbone, such as X101FPN, on the Detectron2 Model Zoo would be beneficial. However, in this section, we will keep this simple backbone model and experiment with the settings and types of optimizers. These are standard hyperparameters because they apply to deep learning in general.

Understanding the available optimizers and their related hyperparameters is essential to understanding the configuration parameters Detectron2 offers for fine-tuning models. This section briefly covers common types of optimizers in deep learning.

Gradient descent

Gradient descent is one of the most essential concepts when it comes to training deep neural networks. Typically, while training neural networks, there is a cost (loss) function (L) that we need to optimize (minimize) concerning the model parameters (W). We use calculus to perform this optimization by getting the derivative of the loss function for the parameters, setting them to zero, and getting the optimal results. However, in complicated deep neural networks (as they often are), it is hard (or even impossible) to solve this equation. Therefore, it is more appropriate to use numerical approximations. *Figure 6.4* illustrates this idea using a simple loss function with one parameter:

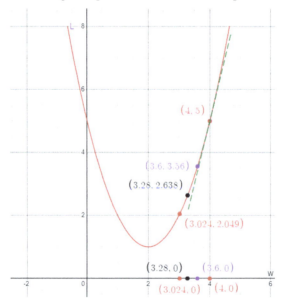

Figure 6.4: Example of gradient descent steps

Specifically, the loss function is $L(w) = w^2 - 4w + 5$, and its derivative is $\nabla L(w) = 2w - 4$. The gradient descent algorithm for this case is simple:

```
Initialize w with a random value
for e in range(num_epochs):
        update w = w - η∇L(w)
```

Here, num_epochs is the number of steps to perform training, and the learning rate (η) dictates how much the parameter, w, updates its value at each training step concerning the derivative of the loss function. These are the two essential hyperparameters of the gradient descent algorithm.

For instance, the current parameter, w, is 4, and the selected learning rate is $\eta = 0.1$. The tangent line (dashed line) for this loss function has a slope that is the derivative of $\nabla L(4) = 4$. At this point, w should move to the left of the value – that is, $w = 3.6$. Similarly, the following steps should correspond to $w = 3.28, 3.024$, and so on. If we keep doing this, the loss function will eventually converge to the optimal point, given the step size is small enough. Notably, when the magnitude of the derivative of the loss function becomes small, the step size becomes small too.

At this point, the question is how to select the number of epochs to train and the learning rate. Generally, choosing a small learning rate makes the learning process slow, and it takes a long time for the model to converge. On the other hand, having a significant learning rate may lead to issues such as the w parameter flickering left and right and not converging.

Let's take a simple code approach to demonstrate these ideas. The following code snippet defines the loss function (calculated on the training data) and its corresponding derivative:

```
def L(w):
    return w**2 - 4*w + 5
def dL(w):
    return 2*w - 4
```

The following code snippet prepares a simple function to simulate the training process with the initial value for the training weight (w0), learning rate (lr), and number of epochs (num_epochs):

```
def train(w0, lr, num_epochs):
    ws, Ls = [], [], w0
    for e in range(num_epochs):
        Ls.append(L(w)) # loss
        ws.append(w) # weight
        dLw = dL(w) # gradient
        w = w - lr*dLw  # update
    return ws, Ls
```

This function returns a list of the weights over time and their corresponding losses. The following code snippet then receives these two lists and visualizes the changes over the training process:

```
# import statements for matplotlib and numpy
def visualize(ws, Ls):
  w_range = np.linspace(-0.5, 4.5, 100)
  L_range = [L(w) for w in w_range]
  plt.plot(w_range, L_range)
  plt.scatter(ws, Ls)
  plt.plot(ws, Ls)
  plt.scatter(ws, [0]*len(ws))
  ax = plt.gca()
  msg = f'lr={lr}\nnum_epochs={num_epochs}'
  plt.text(0.6, 0.95, msg,
      horizontalalignment='left',
      verticalalignment='top',
      transform = ax.transAxes)
  plt.show()
```

Finally, the following code snippet produces the experiment for a specific set of hyperparameters (assuming w0=4.0):

```
num_epochs = 20
lr = 0.1
ws, Ls = train(w0, lr, num_epochs)
visualize(ws, Ls)
```

Figure 6.5 shows the changes in the weights (X-axis) and their corresponding losses (Y-axis) over the training time for different sets of hyperparameters:

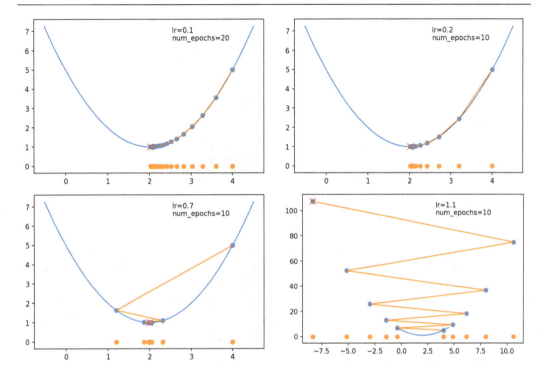

Figure 6.5: Changes in weights (X-axis) and their corresponding losses (Y-axis) over training time for different sets of hyperparameters – learning rate (lr) and the number of epochs (num_epochs)

As we can see, when the learning rate is low (lr=0.1), it takes longer to converge (num_epochs=20) and vice versa (lr=0.2 and num_epochs=10). However, when the learning rate is relatively high (lr=0.7), the changes in the weights flicker left and right several times before converging. Finally, when the learning rate is high (lr=1.1), the gradients explode, and the loss does not converge.

The gradient descent algorithm is robust and straightforward. However, it might be slow to converge in many cases, and the training data may be extensive and not fit in memory for us to perform the loss and its respective derivative. This is when stochastic gradient descent comes in handy.

Stochastic gradient descent

Gradient descent feeds the whole training dataset to the loss function to calculate the output (forward), the loss, and the derivative (gradient) before updating the learning parameter (backward). On the other hand, stochastic gradient descent randomly samples for a number (batch size) of training entries at a time before performing the training step. *Figure 6.6* illustrates the difference between these two approaches:

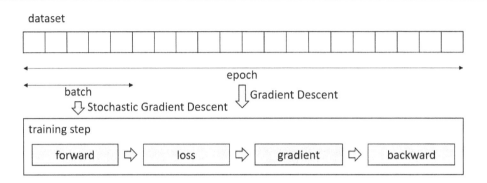

Figure 6.6: Gradient descent versus stochastic gradient descent

Therefore, instead of selecting the number of epochs and learning rate, there is another hyperparameter to select: batch size. The parameter update part of this algorithm remains the same, but the update happens per batch instead of per epoch. Additionally, the random sampling strategy of the batch brings randomness to the training process, which makes this algorithm stochastic. Generally, this algorithm brings two advantages: first, it works when the training dataset is enormous, and second, it brings some randomness and may make the learning process converge faster.

Let's take the code and visualization approach to illustrate these points. The following code snippet imports some packages and prepares a simple dataset for a linear regression task:

```
# some import statements are removed for space efficiency
def yf(x):
    return 2*x - 3 + (np.random.rand(len(x))-0.5)*2

x = np.random.randint(-20, 20, 20)
y = yf(x)

plt.scatter(x, y)
plt.xlabel("x")
plt.ylabel("y")
plt.show()
```

This code snippet produces a typical linear regression dataset with a slope of 2, a bias of -3, and some random noise to make it more realistic. In the future, we assume that we do not know the slope (θ_1) or the bias (θ_2) and try to learn these parameters from this noisy dataset. *Figure 6.7* shows the visualization for this dataset:

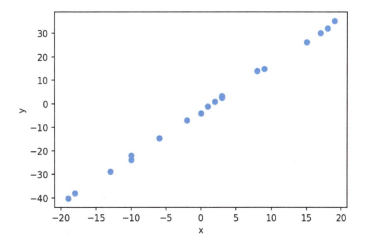

Figure 6.7: A simple dataset for a linear regression task with random noise

The following code snippet prepares the data by combining a bias variable (with values of 1.0) with the existing variable, x, to create an array, X. It also reshapes the y variable so that it becomes a 2D matrix (Y) that supports matrix multiplication:

```
# reshape Y
Y = np.reshape(y, (-1,1))
# add bias
X = np.array(list(zip(x, np.ones_like(x))))
```

Once the data is ready, the following code snippet creates three helper methods to calculate the prediction, the loss, and the gradient for a given set of parameters (theta):

```
# forward
def forward(theta, X):
  Y_hat = np.matmul(X, theta)
  return Y_hat


# loss
def mse(Y_hat, Y):
  return np.mean((Y_hat - Y)**2)


# derrivative
def gradient(X, Y, Y_hat):
  return (2*np.matmul((Y_hat-Y).T, X)/len(Y)).T
```

Notably, the loss method selected is mean squared error (mse). The following code snippet creates an optimizer called GD (gradient descent) to perform parameter updates, given the learning rate and gradients:

```
class GD():
  def __init__(self, lr):
    self.lr = lr

  def backward(self, theta, gradients):
    theta = theta - self.lr*(gradients)
    return theta
```

Next, we must create a function to perform a training step given the X and Y data (either in a batch or the whole epoch):

```
def train_step(X, Y, theta, optimizer):
  # forward
  Y_hat = forward(theta, X)
  loss = mse(Y_hat, Y)
  dLdTheta = gradient(X, Y, Y_hat)
  theta = optimizer.backward(theta, dLdTheta)
  return theta, loss, dLdTheta
```

Now that we know how to perform the training per iteration (either batch or epoch), the following code snippet helps us perform the training loop:

```
def train(X, Y, theta0, optimizer, num_epochs):
  losses = []
  thetas = []
  dLdThetas = []
  theta = theta0
  for epoch in range(num_epochs):
    theta, loss, dLdTheta=train_step(X, Y, theta,optimizer)
    thetas.append(theta)
    losses.append(loss)
    dLdThetas.append(dLdTheta)
  return losses, thetas, dLdThetas
```

Additionally, the following code snippet creates a function that helps us visualize how the losses change over training iterations:

```python
def plot_losses(losses):
  plt.plot(losses)
  plt.ylim([-1, 20])
  ax = plt.gca()
  msg = f'Final loss: {round(losses[-1], 3)}'
  plt.text(0.99, 0.1, msg,
      horizontalalignment='right',
      verticalalignment='top',
      transform = ax.transAxes)
  plt.xlabel("iteration")
  plt.ylabel("loss")
  plt.show()
```

Furthermore, the following code snippet also visualizes how the training parameters (thetas) change over the training iterations:

```python
def plot_thetas(thetas, losses):
  grid_size = 100
  thetas = np.array(thetas)
  theta1s = np.linspace(0, 4, grid_size)
  theta2s = np.linspace(1, -7, grid_size)
  Z = np.zeros((grid_size, grid_size))
  for i in range(grid_size):
    for j in range(grid_size):
      theta = np.array([[theta1s[i]], [theta2s[j]]])
      Y_hat = forward(theta, X)
      loss = mse(Y_hat, Y)
      Z[i,j] = loss
  # some simple plotting code lines are removed for space eff.
```

Next, we must randomly initialize values for the training parameters and set some training hyperparameters:

```python
theta0 = np.random.rand(2, 1)
num_epochs = 500
lr = 0.001
```

The following code snippet performs the training and visualizes the results using the gradient descent approach:

```
optimizer = GD(lr = lr)
losses, thetas, dLdThetas = train(X, Y, theta0,
optimizer=optimizer, num_epochs=num_epochs)
plot_losses(losses)
plot_thetas(thetas, losses)
```

Figure 6.8 shows the losses and thetas over training iterations while using the gradient descent approach:

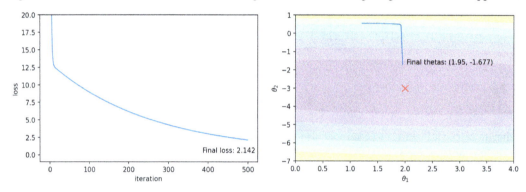

Figure 6.8: The losses and thetas over training iterations when using gradient descent

Observably, the losses and thetas change gradually and smoothly because the training step is performed on the whole dataset. However, the training process does not converge after 500 iterations (the X mark indicates the convergence point).

On the other hand, the following code snippet prepares a method to perform the training process using the stochastic gradient descent approach:

```
def train_mini_batch(X, Y, theta0, optimizer, num_epochs,
batch_size):
    # some simple initialization code lines are removed here
    for e in range(num_epochs):
        # shuffle
        np.random.shuffle(indices)
        for i in range(batches):
            start = i*batch_size
            stop = start + batch_size
            indices_b = indices[start:stop]
```

```
    X_b = X[indices_b]
    Y_b = Y[indices_b]
    theta,loss,dLdTheta=train_step(X_b, Y_b,
                             theta,optimizer)
    thetas.append(theta)
    losses.append(loss)
    dLdThetas.append(dLdTheta)
  return losses, thetas, dLdThetas
```

The following code snippet performs the stochastic gradient descent training and visualizes the results:

```
losses, thetas, dLdThetas = train_mini_batch(X, Y, theta0,
optimizer=optimizer, num_epochs=num_epochs-300, batch_size = 5)
plot_losses(losses)
plot_thetas(thetas, losses)
```

Notably, we reduce the number of epochs by 300. *Figure 6.9* shows the output of this code snippet:

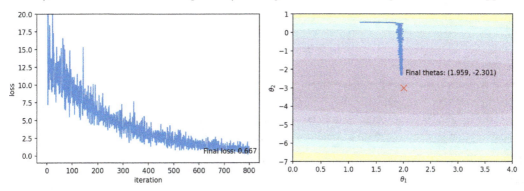

Figure 6.9: The losses and thetas over training iterations when using stochastic gradient descent

As we can see, the iteration is now a batch, not an epoch, and the losses and thetas fluctuate over training iterations due to the randomness introduced by the batch sampling strategy. Even though the number of epochs reduces significantly (200 instead of 500), the final loss and theta values indicate better training.

Momentum

Gradient descent updates the weight based on the value of the current derivative of the loss function only. On the other hand, the momentum approach also considers the accumulation of the past gradients besides the impact of the current gradient. Analogically, we can consider the stochastic/gradient descent approach as pushing a box on the road. If we push, it will move, and if we stop pushing, it will not move due to having high friction. On the other hand, we can consider the momentum approach as pedaling a bicycle. If we pedal, then the bike moves. However, it moves for a bit more when we stop pedaling, thanks to the momentum generated from previous pedalings.

Let's rewrite the stochastic/gradient descent approach weight update formula so that it looks like this:

$$\theta_t = \theta_{t-1} - \eta g_t$$

Here, θ_t and g_t are the training parameter and the gradient at time t, and η is the learning rate. Therefore, if there is no gradient ($g_t = 0$), then the parameter isn't updated. On the other hand, the momentum approach takes two steps. At every iteration, it records a variable, v_t (velocity), and then uses it to update the training parameter using the following two equations:

$$v_t = \mu v_{t-1} + \eta g_t$$

$$\theta_t = \theta_{t-1} - v_t$$

Here, μ typically has a value that is less than and closer to 1.0, such as 0.9, 0.95, or 0.99. One of these values is chosen for μ so that when $g_t = 0$, after a few iterations, v_t also becomes zero and will not update the weight. As these formulas indicate, generally, the main advantage of using the momentum approach is that you can speed up the training time (time to convergence).

Momentum has several variations, and the previously listed one is Sutskever's implementation. There is another popular implementation from Nesterov, which is a little different:

$$v_t = \mu v_{t-1} + g_t$$

$$\theta_t = \theta_{t-1} - \eta v_t$$

Let's continue with the code and visualization approach we introduced in the previous sections to illustrate these concepts. The following code snippet prepares a momentum optimizer:

```python
class Momentum():
    def __init__(self, lr, moment, nesterov=False):
        self.lr = lr
        self.moment = moment
        self.first_step = True
```

```
    self.nesterov = nesterov

  def backward(self, theta, gradients):
    if self.first_step:
      self.v = np.zeros_like(theta)
      self.first_step = False

    if not self.nesterov:
      self.v = self.moment*self.v + self.lr * gradients
      theta = theta - self.v
    else:
      self.v = self.moment*self.v + gradients
      theta = theta - self.lr * self.v

    return theta
```

The following code snippet then performs the training using momentum and visualizes the results:

```
optimizer = Momentum(lr = lr, moment=0.9)
losses, thetas, dLdThetas = train_mini_batch(X, Y, theta0,
optimizer=optimizer, num_epochs=num_epochs-300, batch_size = 5)
plot_losses(losses)
plot_thetas(thetas, losses)
```

Figure 6.10 shows the output of this code snippet. As we can see, the loss and training parameter changes are smoother thanks to the momentum. Additionally, the final loss and theta values indicate that this approach performs even better than the stochastic gradient descent approach without momentum. Nesterov's approach to implementing momentum has similar results. Therefore, we haven't included the code and results here for the sake of space (you can refer to the code file in this book's GitHub repository for these):

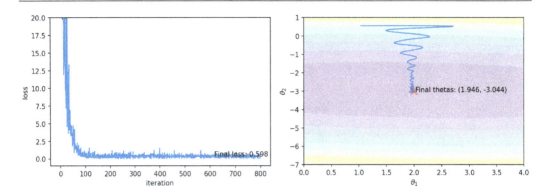

Figure 6.10: The losses and thetas over training iterations when using momentum

In many cases, having a high learning rate when the neural network is not trained well yet may lead to the gradient exploding issue. Likewise, it is better to have a small learning rate when you're getting close to the optimal point so that the changes are not flickering left and right of the optimal point. However, this does not mean we should always select small learning rates since this makes the training process longer. Therefore, a variable learning rate approach may produce better results.

Variable learning rates

There are various ways to approach variable learning rates. Due to space, this section covers the approaches that Detectron2 uses together with its implementation of stochastic gradient descent. Specifically, at the start of training, there is an option to perform a warming-up step. This step takes a hyperparameter called `warmup_iters`. From the start of training, the learning rate gradually increases linearly (or the sinewave approach is also available) so that it becomes the base learning rate at the end of this period.

Additionally, there is another set of hyperparameters called `steps` and `gamma`. These are the steps at which the learning rate changes by `gamma`. These hyperparameters help gradually reduce the learning toward the end of the training process. The following code snippet implements the stochastic gradient descent approach with variable learning rates:

```
class VariableLR():
  def __init__(self, lr, moment,
    nesterov =False,
    warmup_iters  = 0,
    steps    = None,
    gamma    = 0.5):
    # codes removed: self.<inputs> = <inputs>
    self.iter = 0
```

```
    self.lrs = []
    self.first_step = True
def backward(self, theta, gradients):
    # Explained next
```

This class has two methods: `__init__` and `backward`. The first method simply takes the inputs and assigns them to corresponding members of the class. The second one performs the actual updates to the existing parameters (`theta`) using the input gradients (`gradients`):

```
if self.iter in self.steps:
  self.lr = self.lr*self.gamma
self.iter += 1
if self.warmup_iters!=0 and self.iter < self.warmup_iters:
  lr = (self.iter/self.warmup_iters)*self.lr
else:
  lr = self.lr
self.lrs.append(lr)
if self.first_step:
  self.v = np.zeros_like(theta)
  self.first_step = False
if not self.nesterov:
  self.v = self.moment*self.v + lr * gradients
  theta = theta - self.v
else:
  self.v = self.moment*self.v + gradients
  theta = theta - lr * self.v
return theta
```

The following code snippet performs the training and visualizes the results using this approach:

```
optimizer = VariableLR(lr = lr, moment=0.9, nesterov=False,
warmup_iters=50, steps=(300, 500))
losses, thetas, dLdThetas = train_mini_batch(X, Y, theta0,
optimizer=optimizer, num_epochs=num_epochs-300, batch_size = 5)
plot_losses(losses)
plot_thetas(thetas, losses)
```

Figure 6.11 depicts how the learning rate changes over the training iterations:

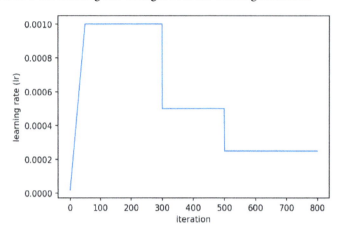

Figure 6.11: Variable learning rates over the training iterations

Figure 6.12 illustrates how the losses and the training parameters change over the training iterations:

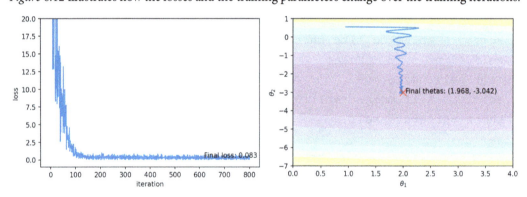

Figure 6.12: The losses and thetas over training iterations when using variable learning rates

As we can see, the variable learning rate approach helps improve performance (it improves the final loss).

There are other popular approaches for implementing variable learning rates, such as the Adam optimization algorithm. However, in practice, stochastic gradient descent often produces similar results. Detectron2 uses stochastic gradient descent, along with options for incorporating momentum and variable learning rates, as the default solver. Therefore, we will not cover other optimization approaches here.

Congratulations! You should have enough theoretical background to understand Detectron2's configurations for its default solver. The following section covers Detectron2's configuration parameters, which are used for training optimization.

Fine-tuning the learning rate and batch size

Detectron2 provides several configurations for its default stochastic gradient descent solver (optimizer). However, these are the main parameters:

- The cfg.SOLVER.IMS_PER_BATCH hyperparameter sets the batch size or the number of images per training iteration.
- The cfg.SOLVER.BASE_LR hyperparameter is used to set the base learning rate.
- The cfg.SOLVER.MOMENTUM hyperparameter stores the momentum value.
- The cfg.SOLVER.NESTROV hyperparameter dictates whether to use Nesterov's implementation of momentum.
- The cfg.SOLVER.WARMUP_ITERS hyperparameter stores the number of warm-up iterations.
- The cfg.SOLVER.STEPS hyperparameter sets the iterations at which the learning rate is reduced by cfg.SOLVER.GAMMA (another hyperparameter).
- The cfg.SOLVER.MAX_ITER hyperparameter sets the maximum number of iterations for training. This iteration is not counted as epochs but as batches.

The following code snippet sets these parameters with the values after we analyze the previously trained results using TensorBoard:

```
cfg.SOLVER.IMS_PER_BATCH = 6
cfg.SOLVER.BASE_LR = 0.001
cfg.SOLVER.WARMUP_ITERS = 1000
cfg.SOLVER.MOMENTUM = 0.9
cfg.SOLVER.STEPS = (3000, 4000)
cfg.SOLVER.GAMMA = 0.5
cfg.SOLVER.NESTROV = False
cfg.SOLVER.MAX_ITER = 5000
```

TensorBoard's inspection of the previous training loss shows a large fluctuation in the losses. Therefore, increasing the batch size would help smoothen the changes. However, we have set this number to 6 due to having limited GPU resources. We also set a higher base learning rate and steps to reduce the learning rate by half (GAMMA).

Congratulations! By now, you should have the necessary background knowledge and code to produce sets of pixel means and standard deviations that are appropriate for your custom images.

Summary

This chapter covered the necessary background knowledge and standard hyperparameters to help you fine-tune Detectron2's solvers. Specifically, it showed you how to use TensorBoard to analyze training results and find insights. Then, we utilized the code and visualization approach to illustrate and find appropriate hyperparameters for Detectron2's solver (optimizer). This chapter also provided a set of hyperparameters deemed suitable for the Detectron2 object detection model, which was trained on a brain tumor dataset. As an exercise, use all the configurations produced in this chapter, perform training experiments, load the results into TensorBoard, and analyze the differences and how these configurations improve accuracy.

This chapter covered the standard set of techniques and hyperparameters since they can be used to fine-tune machine learning in general. The following three chapters will cover fine-tuning techniques for fine-tuning object detection models.

7

Fine-Tuning Object Detection Models

Detectron2 utilizes the concepts of anchors to improve its object detection accuracy by allowing object detection models to predict from a set of anchors instead of from scratch. The set of anchors has various sizes and ratios to reflect the shapes of the objects to be detected. Detectron2 uses two sets of hyperparameters called sizes and ratios to generate the initial set of anchors. Therefore, this chapter explains how Detectron2 processes its inputs and provides code to analyze the ground-truth boxes from a training dataset and find appropriate values for these anchor sizes and ratios.

Additionally, input image pixels' means and standard deviations are crucial in training Detectron2 models. Specifically, Detectron2 uses these values to normalize the input images during training. Calculating these hyperparameters over the whole dataset at once is often impossible for large datasets. Therefore, this chapter provides the code to calculate these values from training batches in a rolling manner.

By the end of this chapter, you will understand how Detectron2 preprocesses its input data. You will also understand the code to generate appropriate values for anchor settings and proper means and standard deviations of pixels for Detectron2 to normalize its input. Finally, this chapter also puts all the configurations derived in the previous chapter and this chapter to train a model and analyze the performance improvements. Specifically, this chapter covers the following topics:

- Setting anchor sizes and anchor ratios
- Setting pixel means and standard deviations
- Putting it all together

Technical requirements

You should have completed *Chapter 1* to have an appropriate development environment for Detectron2. All the code, datasets, and results are available on the GitHub repo of the book at `https://github.com/PacktPublishing/Hands-On-Computer-Vision-with-Detectron2`. It is highly recommended to download the code and follow along.

> **Important note**
> This chapter has code that includes random number generators. Therefore, several values produced in this chapter may differ from run to run. However, the output values should be similar, and the main concepts remain the same.

Setting anchor sizes and anchor ratios

Detectron2 implements Faster R-CNN for object detection tasks, and Faster R-CNN makes excellent use of anchors to allow the object detection model to predict from a fixed set of image patches instead of detecting them from scratch. Anchors have different sizes and ratios to accommodate the fact that the detecting objects are of different shapes. In other words, having a set of anchors closer to the conditions of the to-be-detected things would improve the prediction performance and training time.

Therefore, the following sections cover the steps to (1) explore how Detectron2 prepares the image data for images, (2) get a sample of data for some pre-defined iterations and extract the ground-truth bounding boxes from the sampled data, and finally, (3) utilize clustering and genetic algorithms to find the best set of sizes and ratios for training.

Preprocessing input images

We need to know the sizes and ratios of the ground-truth boxes in the training dataset to find appropriate initial values for them. However, input images may have different sizes and go through various augmentations before feeding to the backbone network to extract features. The configuration values for these hyperparameters are set as if the input images have a standard input size. Therefore, we first need to understand how Detectron2 processes the input images before feeding them to the backbone network.

Therefore, we first use the code snippets provided previously to load the dataset, install Detectron2, register the training dataset, get a training configuration (with `IMS_PER_BATCH = 6`), and get a default trainer. Then, the following code snippet displays the configuration hyperparameters for the image resizing augmentations used by Detectron2:

```
print(cfg.INPUT.MIN_SIZE_TRAIN)
print(cfg.INPUT.MAX_SIZE_TRAIN)
```

This code snippet should display the following output:

```
MIN_SIZE_TRAIN (640, 672, 704, 736, 768, 800)
MAX_SIZE_TRAIN 1333
```

The `ResizeShortestEdge` method uses these hyperparameters to resize input images. Precisely, it resizes shorter edges to one random value selected from `MIN_SIZE_TRAIN` as long as the longer edges are smaller than `MAX_SIZE_TRAIN`. Otherwise, it continues to resize the longer edges to `MAX_SIZE_TRAIN`.

While resizing, the image aspect ratio remains the same. The following code snippet samples for one training batch and displays the sizes of the sampled/transformed images:

```
trainer._data_loader_iter = iter(trainer.data_loader)
data = next(trainer._data_loader_iter)
for i in data:
  print(i['image'].shape)
```

This code snippet has the following output (which may vary in different runs):

```
torch.Size([3, 640, 640])
torch.Size([3, 914, 800])
torch.Size([3, 672, 672])
torch.Size([3, 805, 704])
torch.Size([3, 704, 704])
torch.Size([3, 640, 640])
```

Observably, all the shorter edges are scaled to one of the `MIN_SIZE_TRAIN` values, and none of the longer edges exceed `MAX_SIZE_TRAIN`. Note that these scaled images still have different sizes. Therefore, Detectron2 also performs some more preprocessing using the following code snippet:

```
images = trainer.model.preprocess_image(data)
print(images.tensor.shape)
```

The output of this code snippet should give one image size (which is the largest one rounded to the closest multiple of 32). The following code snippet displays three of the preprocessed images:

```
# some import statements are removed for space efficiency
for i, item in enumerate(images.tensor[:3]):
  img = np.moveaxis(item.to("cpu").numpy(), 0, -1)
  pixel_mean = cfg.MODEL.PIXEL_MEAN
  pixel_std = cfg.MODEL.PIXEL_STD
  img = (img * pixel_std) + pixel_mean
```

```
v = Visualizer(img, metadata={}, scale=0.5)
gt_boxes = data[i]['instances'].get('gt_boxes')
v = v.overlay_instances(boxes = gt_boxes)
dpi = 80
im_data = v.get_image()[:,:, ::-1]
height, width, depth = im_data.shape
figsize = width / float(dpi), height / float(dpi)
fig = plt.figure(figsize=figsize)
plt.imshow(im_data)
plt.imshow(im_data)
plt.axis("off")
plt.show()
```

Figure 7.1 gives examples of images as the output of this code snippet:

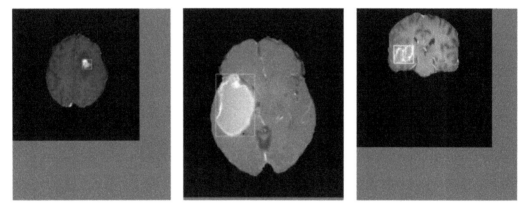

Figure 7.1: Examples of preprocessed images and labels before feeding to the backbone

Observably, smaller images in a batch are padded (bottom and right) to have the maximum size, and all images have the same size. Additionally, bounding boxes are also scaled accordingly.

After understanding how Detectron2 preprocesses input images and ground-truth boxes, we can now sample for several iterations, get the ground-truth bounding boxes from this set, and analyze them to get the right sizes and ratios for training.

Sampling training data and generating the default anchors

The following code snippet samples 1,000 batches (iterations) with a batch size of 6 for images and corresponding ground-truth boxes from the training dataset:

```
from tqdm import tqdm
import torch
def get_gt_boxes_batch(data):
  gt_boxes = [
      item['instances'].get('gt_boxes').tensor
      for item in data
      ]
  return torch.concatenate(gt_boxes)

def get_gt_boxes(trainer, iterations):
  trainer._data_loader_iter = iter(trainer.data_loader)
  gt_boxes = [
      get_gt_boxes_batch(next(trainer._data_loader_iter))
      for _ in tqdm(range(iterations))
      ]
  return torch.concatenate(gt_boxes)

gt_boxes = get_gt_boxes(trainer, 1000)
print()
print(gt_boxes.shape)
```

This code snippet should give us a tensor of bounding boxes. Each bounding box is in the format (x1, y1, x2, y2) for the top-left and lower-right corners of the bounding box. While analyzing the anchors and the bounding boxes, we use width and height information rather than minimum and maximum coordinates. Therefore, the following code snippet converts the coordinates into width and height correspondingly and quickly looks at their minimum and maximum edge sizes:

```
def boxes2wh(boxes):
  x1y1 = boxes[:, :2]
  x2y2 = boxes[:, 2:]
  return x2y2 - x1y1

gt_wh = boxes2wh(gt_boxes)
print(gt_wh.min(), gt_wh.max())
```

This code snippet should have the minimum and maximum edge sizes of all the ground-truth boxes (the box sizes should range from 20 to 300 pixels).

As discussed in *Chapter 3*, Detectron2 has a set of configuration hyperparameters for generating anchors. This code snippet displays these parameters:

```
print("sizes", cfg.MODEL.ANCHOR_GENERATOR.SIZES)
print("ratios", cfg.MODEL.ANCHOR_GENERATOR.ASPECT_RATIOS)
```

This code snippet should display the following output:

```
sizes [[32], [64], [128], [256], [512]]
ratios [[0.5, 1.0, 2.0]]
```

We use a **Feature Pyramid Network** (**FPN**) as the backbone, so the anchor sizes are for different layers. That explains the second dimension for the array of anchor sizes. If we do not use FPN, the size should look like [32, 64, 128, 256, 512] instead. The following code snippet uses Detectron2's default anchor generator to generate a set of anchors from the previous default sets of sizes and ratios and converts them into widths and heights correspondingly:

```
generate_anchors = trainer.model.proposal_generator.anchor_
generator.generate_cell_anchors
anchors = generate_anchors()
ac_wh = boxes2wh(anchors)
```

After having the ground-truth boxes and methods to generate anchors, we are ready to analyze the training bounding boxes and propose a set of initial sizes and ratios that better fit the training data.

Generating sizes and ratios hyperparameters

The ground-truth boxes and anchors are currently in the width and height format, while we need to analyze them in sizes and ratios. Therefore, the following code snippet creates two methods and uses them to convert ground-truth widths and heights into sizes and ratios:

```
def wh2size(wh):
    return torch.sqrt(gt_wh[:, 0]*gt_wh[:, 1])

def wh2ratio(wh):
    return wh[:, 1]/wh[:,0]

gt_sizes = wh2size(gt_wh)
gt_ratios = wh2ratio(gt_wh)
```

Generally, the algorithm to get the best set of sizes and ratios for training has the following steps:

1. Quantify how well each ground-truth box matches a set of anchors (compute best ratios).

2. Quantify the fitness between the set of ground-truth boxes, as a whole, and the set of anchors (compute a fitness score).

3. Perform clustering algorithms on the ground-truth sizes and ratios and get the initializations for these two sets of parameters.

4. Use a genetic algorithm to randomly evolve and get a good set of sizes and ratios for the training process.

Compute the best ratios

For each ground-truth bounding box, there must be a method to quantify how well it can fit with the set of given anchors. We divide the width and height of the ground-truth box by each anchor's width and height and take the inversion. Taking the inversion is necessary because these two boxes' differences are symmetric (two-way). For each anchor, we take the worst ratio to indicate fitness (i.e., if there are suitable matches and bad matches, we will take the worst match as the measure of fitness).

On the other hand, for each ground-truth box, we do not need all the anchors to match well, and it is enough to have at least one anchor that fits well with it. Therefore, we take the maximum ratio among the ratios generated for all anchors as a fitness ratio. This ratio is the indicator of the fitness of this one ground-truth box with the set of anchors. *Figure 7.2* illustrates the steps to compute how well a ground-truth box (gt1) matches two anchors (ac1 and ac2).

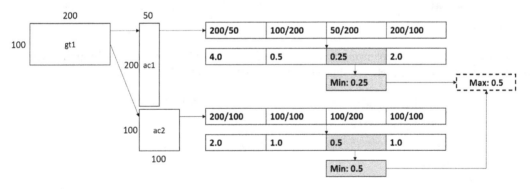

Figure 7.2: Compute how well a ground-truth box (gt1) matches with two anchors (ac1 and ac2)

The following code snippet implements this algorithm:

```
def best_ratio(ac_wh, gt_wh):
    all_ratios = gt_wh[:, None] / ac_wh[None]
    inverse_ratios = 1/all_ratios
```

```
ratios = torch.min(
    all_ratios, inverse_ratios
)
worst = ratios.min(-1).values
best = worst.max(-1).values
return best
```

This function helps quantify the fitness of individual ground-truth boxes with a set of anchors and returns a list of fitness ratios. However, we need one number to quantify the fitness of all ground-truth boxes as a whole, given a set of anchors.

Quantify the fitness

One approach is to take the average of all the fitness scores for all the ground-truth boxes as one fitness score between a set of ground-truth boxes as a whole, given a set of anchors. However, a common practice is to discard the bad matches and not let them contribute to the average calculation. Precisely, we use a threshold, EDGE_RATIO_THRESHOLD, and discard all bad matches.

The default threshold value is 0.25, which means that if the ratio of any side of two boxes differs by more than four times, they are considered bad matches:

```
def fitness(ac_wh, gt_wh, EDGE_RATIO_THRESHOLD = 0.25):
    ratio = best_ratio(ac_wh, gt_wh)
    ret = (ratio*(ratio>EDGE_RATIO_THRESHOLD).float()).mean()
    return ret
```

Additionally, we can use the following function to check the possibility of recalling all the ground-truth boxes using a given set of anchors using a pre-defined EDGE_RATIO_THRESHOLD:

```
def best_recall(ac_wh, gt_wh, EDGE_RATIO_THRESHOLD = 0.25):
    ratio = best_ratio(ac_wh, gt_wh)
    best = (ratio > EDGE_RATIO_THRESHOLD).float().mean()
    return best
```

Specifically, a ground-truth box with edges that do not exceed four times or are not smaller than four times the size of at least one anchor is then recallable.

The following code snippet lets us check whether the default set of sizes and ratios given by Detectron2 can create a good set of anchors that help recall all the ground-truth boxes and how much the fitness score is between these two sets:

```
print("recall", best_recall(ac_wh, gt_wh))
print("fitness", fitness(ac_wh, gt_wh))
```

This code snippet should display the following output:

```
recall tensor(1.)
fitness tensor(0.7838)
```

These results indicate that this default set of hyperparameters is relatively good for this case. However, it is still possible to use a genetic algorithm to improve fitness. The following section uses clustering algorithms to select the initial set of sizes and ratios for the genetic algorithm.

Using clustering algorithms

Clustering the ground-truth sizes and ratios helps to set good initial values for these two sets. The following code snippet prepares two helper methods for clustering a set of values and visualizing the results:

```python
import numpy as np
from scipy.cluster.vq import kmeans
def estimate_clusters(values, num_clusters, iter=100):
  std = values.std(0).item()
  k, _ = kmeans(values / std, num_clusters, iter=iter)
  k *= std
  return k

def visualize_clusters(values, centers):
  plt.hist(values, histtype='step')
  plt.scatter(centers, [0]*len(centers), c= "red")
  plt.show()
```

The following code snippet then utilizes these two methods to cluster and visualize the ground-truth sizes and ratios:

```python
sizes = estimate_clusters(gt_sizes, 5)
visualize_clusters(gt_sizes, sizes)
ratios = estimate_clusters(gt_ratios, 3)
visualize_clusters(gt_ratios, ratios)
```

Notably, we use five sizes and three ratios. Therefore, we set these values as the number of clusters in these two cases. *Figure 7.3* shows the histograms of the ground-truth sizes and ratios, and the dots represent the centers of their clusters.

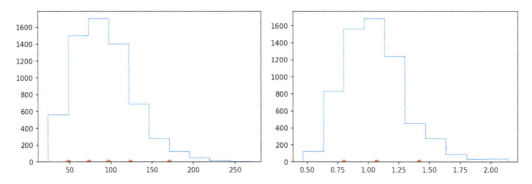

Figure 7.3: Histograms of sizes and ratios and the centers of their clusters

Most sizes are 100 pixels, and the ratios are more toward 1.0 because most boxes are square. Let us run the following code snippet to see how well this new set of sizes and ratios fits our ground-truth boxes:

```
fitness(boxes2wh(generate_anchors(sizes, ratios)), gt_wh)
```

This snippet should print some value like the following:

```
tensor(0.8786, dtype=torch.float64)
```

This score is a much better fit than the default anchor set (0.7838). However, there is still an opportunity to improve this score using a genetic algorithm.

Evolve the results using the genetic algorithm

The genetic algorithm for this case is straightforward. *Figure 7.4* illustrates the main steps of this algorithm, with five steps to perform iteratively for a selected number of iterations.

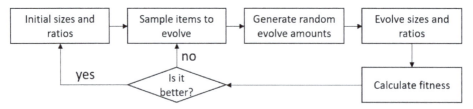

Figure 7.4: Flowchart of the genetic algorithm for generating sizes and ratios

This algorithm has the following steps:

1. Start with the initial sizes and ratios generated by the clustering algorithms.
2. Sample some items to evolve based on a given probability (e.g., 0.9 to have an item evolve 90% of the time).
3. Sample some random evolving amounts from a normal random distribution with a given mean and standard deviation.
4. Change the sizes and ratios using the generated random amounts.
5. Compute a fitness score using the new sets of sizes and ratios.
6. If the fitness improves, set the initial set of sizes and ratios to these current sets and loop back to *step 1*. Otherwise, proceed to *step 2*.

The following code snippet implements this algorithm:

```python
def evolve(sizes, ratios,
           gt_wh,
           iterations   = 10000,
           probability  = 0.9,
           muy          = 1,
           sigma        = 0.05,
           fit_fn       = fitness,
           verbose      = False):
  # Step 1: Current anchors and their fitness
  msg = f"Evolving ratios and sizes:"
  pbar = tqdm(range(iterations), desc = msg)
  for i, _ in enumerate(pbar):
    # Step 2 and 3: sample evolving items and amounts
    # Step 4 and 5: change sizes and ratios and new fitness


    if mutated_fit > best_fit:
      # Step 6: update sizes and ratios if better
  return sizes, ratios
```

The code for each step is explained in the following sections. Specifically, the code for *step 1* generates the anchors using the current sizes and ratios and computes the current fitness score:

```
anchors = generate_anchors(tuple(sizes), tuple(ratios))
ac_wh = boxes2wh(anchors)
anchor_shape = len(sizes) + len(ratios)
best_fit = fit_fn(ac_wh, gt_wh)
```

The following code snippet provides the codes for *steps 2* and *3*. These steps randomly sample items to mutate and the amounts for mutation:

```
mutation = np.ones(anchor_shape)
mutate = np.random.random(anchor_shape) < probability
mutation = np.random.normal(muy, sigma, anchor_shape)*mutate
mutation = mutation.clip(0.3, 3.0)
```

Steps 4 and *5* make changes to the sizes and ratios after getting information about what to change and how much to change. Additionally, it also computes the new fitness score after performing mutation. The source code for these steps is as follows:

```
mutated_sizes = sizes.copy()*mutation[:len(sizes)]
mutated_ratios = ratios.copy()*mutation[-len(ratios):]
mutated_anchors = generate_anchors(
    tuple(mutated_sizes),
    tuple(mutated_ratios))
mutated_ac_wh = boxes2wh(mutated_anchors)
mutated_fit = fit_fn(mutated_ac_wh, gt_wh)
```

Lastly, if the new fitness score is better, the code for *step 6* updates the best score, sizes, and ratios with their corresponding current values:

```
sizes = mutated_sizes.copy()
ratios = mutated_ratios.copy()
best_fit = mutated_fit
msg=f"Evolving ratios and sizes, Fitness = {best_fit:.4f}"
pbar.desc = (msg)
```

We can then run the following line of code to start the optimization process:

```
e_sizes,e_ratios = evolve(sizes,ratios,gt_wh,verbose=False)
```

This algorithm should produce a fitness score of approximately 0.8835. This result is slightly better than the sizes and ratios generated by the clustering algorithm. Finally, we should run the following code snippet and save these sizes and ratios for future training:

```
print("sizes", e_sizes)
print("ratios", e_ratios)
```

This code snippet should display the following output (or similar):

```
sizes [ 68.33245953 112.91302277  89.55701886
144.71037342  47.77637482]
ratios [0.99819939 0.78726896 1.23598428]
```

Therefore, the training configuration should use the following hyperparameters for sizes and ratios:

```
cfg.MODEL.ANCHOR_GENERATOR.SIZES = [[68.33245953],
[112.91302277],  [89.55701886], [144.71037342],  [47.77637482]]
cfg.MODEL.ANCHOR_GENERATOR.ASPECT_RATIOS = [[0.99819939,
0.78726896, 1.23598428]]
```

Congratulations! By this time, you should have a deeper understanding of how Detectron2 processes its inputs and how to select a set of default hyperparameters for generating anchors. The following section covers the steps to set another set of essential parameters to fine-tune Detectron2 models. They are the means and standard deviations of the pixel values in the training images.

Setting pixel means and standard deviations

Input image pixels' means and standard deviations are crucial in training Detectron2 models. Specifically, Detectron2 uses these values to normalize the input images. Detectron2 has two configuration parameters for these. They are cfg.MODEL.PIXEL_MEAN and cfg.MODEL.PIXEL_STD. By default, the common values for these two hyperparameters generated from the ImageNet dataset are [103.53, 116.28, 123.675] and [57.375, 57.120, 58.395]. These values are appropriate for most of the color images. However, this specific case has grayscale images with different values for pixel means and standard deviations. Therefore, producing these two sets of values from the training dataset would be beneficial. This task has two main stages: (1) preparing a data loader to load images and (2) creating a class to calculate running means and standard deviations.

Preparing a data loader

Detectron2's data loader is iterable and can yield infinite data batches with random inputs. However, we would like to analyze each of the input images once. Therefore, creating a custom dataset and using Pytorch's data loader to load the input images is more convenient. The following code snippet imports the required packages and sets the initial configurations:

```
# some import statements are removed for space efficiency
name_ds = "braintumors_coco"
af = "_annotations.coco.json"
img_dir = name_ds + "/train/"
json_file_train = name_ds + "/train/" + af
batch_size = 64
num_workers = 2
```

Notably, we do not perform training, so a higher batch size is used (64 in this case) due to having more available GPUs. The following code snippet prepares a PyTorch dataset class for our custom dataset:

```
# some import statements are removed for space efficiency
class TumorDataset(Dataset):
  def __init__(self,
                  data,
                  img_dir="",
                  transform = None):
    self.data = data
    self.img_dir = img_dir
    self.transform = transform
  def __len__(self):
    return len(self.data)

  def __getitem__(self, idx):
    file_name  = os.path.join(self.img_dir,
                              self.data[idx]['file_name'])
    image = cv2.imread(file_name)
    if self.transform is not None:
        image = self.transform(image = image)['image']
    return image
```

The input images have different sizes. Therefore, we transform all the input images into the same size using the albumentations package:

```
import albumentations as A
from albumentations.pytorch import ToTensorV2
image_size = 640
augs = A.Compose([
    A.Resize(height = image_size, width = image_size),
    ToTensorV2()])
```

Notably, we resize all input images to 640 by 640. The following code snippet reads the JSON annotation file and creates a PyTorch data loader:

```
import json
with open(json_file_train) as f:
  ds = json.load(f)['images']

tds = TumorDataset(ds, img_dir=img_dir, transform=augs)
image_loader = DataLoader(tds,
                          batch_size   = batch_size,
                          shuffle      = False,
                          num_workers = num_workers,
                          pin_memory   = True)
```

This data loader loads the data in batches because the input dataset might be too large to fit the memory at one time. Therefore, we need to create a class to calculate the means and standard deviations in a rolling/running manner.

Calculating the running means and standard deviations

The formulas for calculating the mean $(\overline{x})$ and standard deviation (std) of a set of n values, if you have access to the whole set of values, are straightforward:

$$sum = \sum_{i=1}^{n} x_i$$

$$\overline{x} = \frac{sum}{n}$$

$$ssd = \sum_{i=1}^{n} (x_i - \overline{x})^2$$

$$std = \sqrt{\frac{ssd}{n-1}}$$

However, the calculations for these two statistics in a running manner are a little more involved. Therefore, the following formulas may be a little complicated but mostly because of the mathematical notation. Specifically, we must keep track of these statistics and then update them after reading every batch of data. Let $\overline{x}_c$, sum_c, ssd_c, and std_c denote the current statistics with currently seen items (n_c). Now, when we see a new batch of data b with n_b items, the statistics (sum_b and ssd_b) are computable over this batch. The mean ($\overline{x}_n$)of the new dataset should be as follows:

$$\overline{x}_n = \frac{1}{n_c + n_b}(sum_c + sum_b) = \frac{1}{n_c + n_b}(n_c\overline{x}_c + sum_b)$$

All these values are stored and computable from the batch data. Therefore, the new mean is updated accordingly. Similarly, the sum of the squared differences of this new dataset is as follows:

$$ssd_n = \sum_{i=1}^{n_c} (x_i - \overline{x}_n)^2 + \sum_{j=1}^{n_b} (b_j - \overline{x}_n)^2 = \sum_{i=1}^{n_c} (x_i - \overline{x}_n)^2 + ssd_b$$

The sum of squared differences for the batch (ssd_b) can be computed from the current batch data. However, we need to unpack the other part of the computation as the sum of squared differences of the current data items with respect to the new mean, as follows:

$$ssd_{c\_new\_mean} = \sum_{i=1}^{n_c}(x_i - \overline{x}_n)^2 = \sum_{i=1}^{n_c}(x_i)^2 - 2\overline{x}_n\sum_{i=1}^{n_c}x_i + n_c(\overline{x}_n)^2 = \sum_{i=1}^{n_c}(x_i)^2 - 2\overline{x}_n sum_c + n_c(\overline{x}_n)^2$$

Similarly, we can have the sum of squared differences for the list of current items:

$$ssd_c = \sum_{i=1}^{n_c}(x_i - \overline{x}_c)^2 = \sum_{i=1}^{n_c}(x_i)^2 - 2\overline{x}_c\sum_{i=1}^{n_c}x_i + n_c(\overline{x}_c)^2 = \sum_{i=1}^{n_c}(x_i)^2 - 2\overline{x}_c sum_c + n_c(\overline{x}_c)^2$$

Then, the sum of squares of the previously seen items can be written as follows:

$$\sum_{i=1}^{n_c}(x_i)^2 = ssd_c + 2\overline{x}_c sum_c - n_c(\overline{x}_c)^2$$

Now, we can compute the sum of squared differences of the current items with respect to the new mean as follows:

$$ssd_{c\_new\_mean} = ssd_c + 2\overline{x}_c sum_c - n_c(\overline{x}_c)^2 - 2\overline{x}_n sum_c + n_c(\overline{x}_n)^2$$

This equation now contains all computable values. Therefore, we can compute the sum of squared differences for the new set of data (the current one and the new batch) and, thus, the standard deviation.

The following code snippets implement these calculations. First, we need a helper method to broadcast the running means into the sizes compatible with the input sizes used in our calculation:

```
def broad_cast(x, image_size, channel):
  y = torch.broadcast_to(x, (image_size**2, channel))
  z = y.reshape(image_size, image_size, 3)
  return torch.moveaxis(z, 2, 0)
```

The following code snippet utilizes this helper method and prepares a class to calculate running means and standard deviations from batches of inputs in a rolling/running manner:

```
class RunningStats:
  def __init__(self):
    # initialize: self.n = 0, self.mean = 0, self.ssd = 0
  def push(self, x):
    dims = [0, 2, 3]
    count = 1
    for dim in dims:
      count *= x.shape[dim]
    image_size = x.shape[-1]
    channel = x.shape[1]
    if self.n == 0: # start
      # Step 1: compute new_mean, new_ssd, new_count
    else: # otherwise
      # Step 2: store old statistics and compute new ones
    # release results
    self.mean = new_mean
    self.ssd = new_ssd
    self.n = new_count
    self.std = torch.sqrt(new_ssd/(new_count-1))
```

There are two main computation steps in the `push` method of this class. *Step 1* is computed at the start and is straightforward. It computes the statistics using the first batch of data without caring about existing statistics:

```
new_mean = x.sum(axis=dims)/count
new_ssd = ((x - broad_cast(new_mean, image_size, channel))**2).
sum(axis=dims)
new_count = count
```

Step 2 stores the old statistics and computes the new statistics using the formulas explained previously:

```
old_count, old_mean, old_ssd = self.n, self.mean, self.ssd
old_sum = old_mean * old_count
new_count = self.n + count
new_sum = old_sum + x.sum(axis=dims)
new_mean = new_sum/(self.n + count)
old_ssd_new_mean = (
    old_ssd
    + 2*old_mean*old_sum
    - old_count*(old_mean)**2
    - 2*new_mean*old_sum
    + old_count*(new_mean)**2)
new_ssd = (old_ssd_new_mean +
            ((x - broad_cast(new_mean,
                            image_size,
                            channel))**2
        ).sum(axis=dims))
```

Finally, the following code snippet performs the main analysis tasks and displays the results:

```
rs = RunningStats()
for inputs in tqdm(image_loader):
    rs.push(inputs)
print()
print(rs.mean)
print(rs.std)
```

This code snippet should display the following output:

```
tensor([20.1962, 20.1962, 20.1962])
tensor([39.5985, 39.5985, 39.5985])
```

We can then use these values for training configuration as follows:

```
cfg.MODEL.PIXEL_MEAN = [20.1962, 20.1962, 20.1962]
cfg.MODEL.PIXEL_STD = ([39.5985, 39.5985, 39.5985]
```

Congratulations! By this time, you should have some background knowledge and the code to produce sets of pixel means and standard deviations appropriate for your custom images. Next, we should put all the configurations derived in the previous chapter and this chapter and retrain our brain tumor detection model.

Putting it all together

The code for training the custom model with the ability to perform evaluations and a hook to save the best model remains the same as in *Chapter 5*. However, the configuration should be as follows:

```
# Codes to generate cfg object are removed for space effc.
# Solver
cfg.SOLVER.IMS_PER_BATCH = 6
cfg.SOLVER.BASE_LR = 0.001
cfg.SOLVER.WARMUP_ITERS = 1000
cfg.SOLVER.MOMENTUM = 0.9
cfg.SOLVER.STEPS = (3000, 4000)
cfg.SOLVER.GAMMA = 0.5
cfg.SOLVER.NESTROV = False
cfg.SOLVER.MAX_ITER = 5000
# checkpoint
cfg.SOLVER.CHECKPOINT_PERIOD = 500
# anchors
cfg.MODEL.ANCHOR_GENERATOR.SIZES = [[68.33245953,
112.91302277,  89.55701886, 144.71037342,  47.77637482]]
cfg.MODEL.ANCHOR_GENERATOR.ASPECT_RATIOS = [[0.99819939,
0.78726896, 1.23598428]]
# pixels
cfg.MODEL.PIXEL_MEAN = [20.1962, 20.1962, 20.1962]
```

```
cfg.MODEL.PIXEL_STD = [39.5985, 39.5985, 39.5985]
# Other params similar to prev. chapter are removed here
```

Please refer to the complete Jupyter notebook on GitHub for the code and some example output. You can also load the example of the train results and analyze them using TensorBoard. Notably, these few modifications improve the model's accuracy significantly (mAP@0.5 should be approximately 50%) without sacrificing training or inferencing time.

Summary

This chapter provides code and visualizations to explain how Detectron2 preprocesses its inputs. In addition, it provides code to analyze the ground-truth bounding boxes and uses a genetic algorithm to select suitable values for the anchor settings (anchor sizes and ratios). Additionally, it explains the steps to produce the input pixels' means and standard deviations from the training dataset in a running (per batch) manner when the training dataset is large and does not fit in memory at once. Finally, this chapter also puts the configurations derived in the previous chapter and this chapter into training. The results indicate that with a few modifications, the accuracy improves without impacting training or inferencing time. The next chapter utilizes these training configurations and the image augmentation techniques (introduced next) and fine-tunes the Detectron2 model for predicting brain tumors.

8
Image Data Augmentation Techniques

This chapter answers the questions of what, why, and how to perform image augmentation by providing a set of standard and state-of-the-art image augmentation techniques. Once you have foundational knowledge of image augmentation techniques, this chapter will introduce Detectron2's image augmentation system, which has three main components: `Transformation`, `Augmentation`, and `AugInput`. It describes classes in these components and how they work together to perform image augmentation while training Detectron2 models.

By the end of this chapter, you will understand important image augmentation techniques, how they work, and why they help improve model performance. Additionally, you will be able to perform these image augmentations in Detectron2. Specifically, this chapter covers the following topics:

- Image augmentation techniques
- Detectron2's image augmentation system:
  - Transformation classes
  - Augmentation classes
  - The AugInput class

Technical requirements

You must have completed *Chapter 1* to have an appropriate development environment for Detectron2. All the code, datasets, and results are available in this book's GitHub repository at `https://github.com/PacktPublishing/Hands-On-Computer-Vision-with-Detectron2`.

Image augmentation techniques

Image augmentation techniques help greatly improve the robustness and accuracy of computer vision deep learning models. Detectron2 and many other modern computer vision architectures use image augmentation. Therefore, it is essential to understand image augmentation techniques and how Detectron2 uses them. This section covers **what** image augmentations are, **why** they are important, and introduces popular methods to perform them (**how**). The next two sections explain how Detectron2 uses them during training and inferencing.

Why image augmentations?

Deep learning architectures with a small number of weights may not be accurate (**bias issue**). Therefore, modern architectures tend to be complex and have huge numbers of weights. Training these models often involves passing through the training datasets for several epochs; one epoch means the whole training dataset is passed through the model once. Therefore, the huge numbers of weights may mean the models tend to memorize trivial features in the training dataset (**overfitting issue**) and do not work well in future unseen data (**variance issue**).

One popular example that's often used to illustrate this case is a deep learning model trained to detect the difference between Golden Retrievers and Huskies. It showed great accuracy on the training and evaluation sets during training. However, in reality, it performed poorly. Further inspection of the model indicates that it used the snowy background in images to differentiate between a Husky and a Golden Retriever rather than their actual characteristics. Therefore, the detection objects must be placed in different contexts.

One straightforward solution is to have more training data. However, in many cases, having more images may not be possible. For instance, collecting medical images from patients may result in legal restrictions. Even if collecting more images is possible, manual labeling costs in such areas might be huge because it requires highly skilled people to perform labeling. Therefore, image augmentation techniques are important.

What are image augmentations?

Image augmentation techniques transform images and their related annotations (for example, bounding boxes, segments, and key points) and produce slightly different copies. New copies of the same input allow the model to have further training data with different contexts to tackle overfitting and improve its robustness. Thanks to image augmentation, several modern deep learning architectures for computer vision can even produce models that never look at an image more than once during training. The following section covers the most important and useful image augmentation techniques.

How to perform image augmentations

The two most commonly used and straightforward techniques are resize and scale, which are often used to preprocess inputs. Additionally, there is a vast number of image augmentation techniques, and they can be applied at the image level or annotation level (for example, the bounding box). This section only covers the image-level image augmentation techniques, but the ideas are similar to annotation-level ones. These techniques can be broadly categorized into photometric distortions, geometric distortions, and image occlusions.

Photometric distortions

This category of image augmentations includes techniques such as hue, saturation, brightness, exposure, grayscale, blur, and noise. *Figure 8.1* illustrates examples of these techniques:

Figure 8.1: Examples of augmentation techniques with photometric distortions

These techniques change input images for different contexts related to colors. These differences help the vision models focus on learning features from the shapes of the objects to be detected and not just the colors. These techniques do not change the images geometrically; thus, they do not change image labels. Let's look at these in more detail:

- **Hue**: This technique changes the color channel of the input images. In other words, this technique changes the color tones or color schemes of the input images, such as making them look more blue or green, and produces more images.

- **Saturation**: This technique changes how colorful the color channel of the input images is. It is analogous to mixing some amount of white color with the current color. For instance, adding some white amount to a blue color makes it less blue.

- **Brightness**: This technique changes the light conditions of the input images. It is analogous to shedding some amount of light to color. If there is no light, the input becomes black.

- **Exposure**: This technique is similar to brightness. However, exposure is biased to highlight tones, while brightness changes all tones equally.

- **Grayscale**: This technique changes the input image to a single-channel image. In other words, the pixel values typically range from 0 to 255, and it is a single channel, or if it retains three channels, all channels have the same value per pixel location.

- **Blur**: This technique smoothen the images to help the model be more robust to camera focus. For instance, if Gaussian blur is used, a given pixel takes the weighted average of pixels surrounding it (the weights follow a Gaussian distribution).

- **Noise**: This technique adds random noise (typically salt-and-pepper noise) to the image and creates a new version. This technique makes the training models more robust to camera artifacts and tackles adversarial attacks.

Now, let's look at geometric distortions.

Geometric distortions

This category of image augmentations includes techniques such as flip, crop, rotation, and shear. *Figure 8.2* illustrates examples of these techniques:

Figure 8.2: Examples of augmentation techniques in geometric distortions

These techniques change input images to different contexts related to the location and perspectives where the images are collected. These differences in perspectives help provide the vision models with training data containing different views. These techniques change the images geometrically; thus, they also change image annotations such as bounding boxes:

- **Flip**: This technique flips the input image horizontally or vertically and produces a new version of this input. This technique helps make the training model robust to object orientation changes.

- **Crop**: This technique selects a rectangular region in the original image, removes other parts, and produces a new image. This technique is often combined scaling to resize the image to the original size. This technique helps simulate changes in camera position and/or object movements.

- **Rotation**: This technique rotates the original image to a certain degree around the center of the image and creates a new version of the input. This technique helps add images such as those taken with different camera rolls. In other words, it helps train models that are robust to cameras and objects that are not perfectly aligned.

- **Shear**: This technique shears input images horizontally, vertically, or both to add different perspectives to the input image and create new versions. This technique helps add more pictures as if they were taken from different pitches and yaws of the camera.

Next, we'll look at image occlusions.

Image occlusions

This category of image augmentations includes techniques such as Cutout, MixUp, CutMix, and Mosaic. These techniques help the model learn an object as a whole and not just focus on a particular part of it so that when we see that part of an object is occluded, the model can still recognize it. Additionally, it places the object in other contexts (for example, different backgrounds or surrounding objects) so that the model focuses on the features of the object itself instead of the contexts:

| Cutout | Mixup | CutMix | Mosaic |

Figure 8.3: Examples of augmentation techniques in image occlusions

Let's look at these techniques in more detail:

- **Cutout**: This technique removes part of an input image (replacing it with other values such as black or the average color of the images in the dataset) and creates a new version. This technique helps make the training model more robust to the cases where some other objects may cover part of the object. This technique forces the model to learn features from an object as a whole and does not just focus on a particular part of it. This technique may or may not change the annotations of the input image, depending on the cutout regions.

- **MixUp**: This technique blends two input images and their related annotations into one with different ratios and creates a new training data item. It allows us to train models on convex combinations of pairs of input images and their labels. These convex combinations regularize the training model to favor simple linear behavior across training examples.

- **CutMix**: This technique replaces one region of an input image with a part of another image based on a defined ratio and produces new versions. This technique may remove annotations from the input image and include annotations from the other image. This technique works well by leveraging the advantages of both the cutout and the mixup approaches.

- **Mosaic**: This technique combines four images (versus two in CutMix) based on some defined ratios and produces new versions. This technique leverages the advantages of the other image occlusion techniques listed previously. Additionally, it is useful if the mini-batch size is small due to having limited computation resources because one image is similar to four in this case.

Congratulations! By now, you should understand what image augmentation techniques are, why they are important, and how to perform them. The next section covers how to apply image augmentation techniques using Detectron2.

Detectron2's image augmentation system

Detectron2's image augmentation system has three main groups of classes: `Transformation`, `Augmentation`, and `AugInput`. These components help augment images and their related annotations (for example, bounding boxes, segment masks, and key points). Additionally, this system allows you to apply a sequence of declarative augmentation statements and enables augmenting custom data types and custom operations. *Figure 8.4* shows a simplified class diagram of Detectron2's augmentation system:

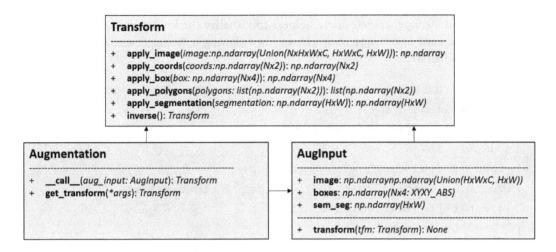

Figure 8.4: Simplified class diagram of Detectron2's augmentation system

The `Transform` and `Augmentation` classes are the bases for all the classes in their respective groups. Notably, the data format for boxes is in `XYXY_ABS` mode, which dictates the boxes to be in (`x_min`, `y_min`, `x_max`, `y_max`), specified in absolute pixels. Generally, subclasses of the `Transform` base class perform the deterministic changes of the input data, while subclasses of the `Augmentation` base class define the policy with some randomness and generate objects of the `Transform` subclasses. The `AugInput` class is used to encapsulate input data for `Augmentation` classes, and it can be used independently (via the `transform` method) or as the input for the `Augmentation` classes. One typical usage of this system should include the following steps:

1. Create an input using the `AugInput` class (`input`).

2. Declare one or more instances of the `Augmentation` class (`augs`).

3. Apply the augmentations to the input (in place) and return an object of the `Transform` class: `transform = augs(input)`.

4. Use the returned `transform` object to perform augmentations on extra data if needed.

First, we must prepare some common code snippets that are used by these sections. First, the following code snippet downloads and unzips some sample images and installs Detectron2:

```
!wget --quiet <url_to_Chapter08.zip>
!unzip --q Chapter08.zip
!python -m  pip install --q \
'git+https://github.com/facebookresearch/detectron2.git'
```

The following code snippets create some helpers to visualize the inputs and outputs of the `Transformation` and `Augmentation` classes. The following is the helper method that shows images in their original size:

```
# some import statements are removed for space efficiency
def imshow(image):
  dpi = plt.rcParams["figure.dpi"]
  im_data = image[:,:, ::-1]
  height, width, depth = im_data.shape
  figsize = width / float(dpi), height / float(dpi)
  fig = plt.figure(figsize=figsize)
  plt.imshow(im_data)
  plt.imshow(im_data)
  plt.axis("off")
  plt.show()
```

The following method (`visualize_image_boxes`) visualizes bounding boxes on input images:

```
def visualize_image_boxes(img, boxes, colors=['k']):
  v = Visualizer(img, metadata={})
  v = v.overlay_instances(
      boxes           = boxes,
      assigned_colors = colors*len(boxes)
      )
  im_data = v.get_image()
  imshow(im_data)
```

Finally, the following method (`visualize_sample`) gets a dictionary (`sample`) as input, loads images and bounding boxes, and visualizes them while leveraging the previously built method:

```
def visualize_sample(sample):
  img = sample['image'].to("cpu").numpy()
  img = np.moveaxis(img, 0, -1)
  boxes = sample['instances'].get('gt_boxes')
  visualize_image_boxes(img, boxes)
```

These methods help visualize the data sample in the image's actual size, along with its corresponding bounding boxes, with data extracted from the Detectron2 dataset.

The following code snippet imports the `transforms` package from Detectron2. This package implements Detectron2's image augmentation system:

```
import detectron2.data.transforms as T
```

In the following sections, we will learn more about the classes in Detectron2's image augmentation system.

> **Important note**
>
> In the following sections, for some classes, there will be examples to illustrate how they work and strengthen the usage patterns of the classes in Detectron2's augmentation system. However, for the sake of space, there are no examples for the classes that are straightforward to understand. Many of these classes need to perform interpolation and have `interp` as an input parameter; this `interp` parameter has the same meaning across different classes. Thus, we will not repeatedly explain this parameter through different classes but merely lists it as an input parameter if it is one.

Transformation classes

A transformation class implements deterministic changes to the input images and related annotations based on some input parameters. One example is to scale an image and its bounding boxes by 0.9; it will always perform this operation in the same way for all inputs, and no randomness is involved. Besides the image, a transformation may change its annotations of different data types, such as bounding boxes, segmentation masks, and key points. Therefore, there are methods for these data types in the form of `apply_*`, where * denotes the data types, such as `image`, `box`, `coords`, and `polygons`.

Detectron2 provides classes to perform commonly used transformations, and users can define a new class that implements custom transformations and methods for different annotation types. The following are the common transformation classes:

- **The Transform class** is the base class for all other transformations to implement and extend. It has abstract method signatures for `apply_image` and `apply_coords` and has built-in implementations for `apply_box` and `apply_polygons` that use `apply_coords`. There is also a built-in implementation for `apply_segmentation` that utilizes `apply_image`. Finally, there is a method signature for the `inverse` method. This method should be implemented to return a `Transform` object that helps inverse the changes made by the original transformation. This method is used for geometric distortions only. The inverse of photometric distortions that do not change coordinates is an instance of the `NoOpTransform` class.

- **The NoOpTransform class** is also a subclass of the `Transform` class. Its instance is used in scenarios such as when returning the `inverse()` method in the transformation classes in the photometric distortions category. It simply does nothing. Specifically, the implementation of the `apply_image` and `apply_coords` abstract methods returns their corresponding inputs, and the implementation of the `inverse` method returns the current object of this class (`self`).

- **The ResizeTransform class** is one of the most commonly used transformations because other transformations and deep learning models often require images to be the same size. In contrast, input images may come in different sizes. This transformation accepts the original image height and width (`h, w`), the expected height and width (`new_h, new_w`), and the interpolation method (`interp`) and performs the resize accordingly.

- **The ExtentTransform class**: This class takes a subregion (`sub_rect`) of an input image (in `XYXY_ABS` format), the expected output size (`h, w`), and the interpolation method (`interp`) and resizes the subtracted region into the expected size using the given interpolation method. Additionally, if any input subregion sizes are larger than the input image, a fill color (the `fill` input parameter) is used for padding. The following code snippet illustrates the usage of this class:

```
# data
img = cv2.imread("Chapter08/1.jpg")
boxes = [[150, 140, 350, 420]]
visualize_image_boxes(img, boxes = boxes, colors=['k'])
# transform
e_tfm = T.ExtentTransform(
    src_rect = (100, 100, 400, 470),
    output_size = img.shape[:2]
    )
et_img = e_tfm.apply_image(img)
et_boxes = e_tfm.apply_box(boxes)
visualize_image_boxes(et_img, boxes = et_boxes)
```

This snippet loads a sample image with a ground-truth box. Then, it creates an `ExtentTransform` object and calls `apply_image` and `apply_box` to transform the input image and ground-truth box accordingly. *Figure 8.5* shows the output of this code snippet:

Original image and ground-truth box   Extended image and ground-truth box

Figure 8.5: Original image and the ground-truth box and their extent-transformed versions

It is relatively easy to create a `Transform` object and consistently apply the transformation to the original images and their corresponding annotations by calling the `apply_*` methods. As we can see, this specific snippet helps zoom the inputs in and train models to detect larger-sized objects.

- **The RotationTransform class**: This class takes the input size of the original image (`h`, `w`), `angle` in degrees for rotation, and rotates the input image and its related annotations around the center (or a point set by the `center` option) of the picture. We can also set the `expand` option to `True` to resize the image so that it fits the whole rotated size (default) or crop it (if set to `False`). The following code snippet shows a typical usage of this transformation:

```
r_tfm = T.RotationTransform(
    img.shape[:2],
    angle = 15,
    expand=False
    )
rt_img = r_tfm.apply_image(img)
rt_boxes = r_tfm.apply_box(boxes)
visualize_image_boxes(rt_img, boxes = rt_boxes)
```

Figure 8.6 (right) illustrates the rotated versions of the original input image and ground-truth box (left):

Original image and ground-truth box   Rotated image and ground-truth box

Figure 8.6: Original image and the ground-truth box and their rotated versions

As we can see, the usage pattern remains the same, with different input parameters for generating appropriate information for the transformation.

- **The PILColorTransform and ColorTransform classes** are the generic wrappers for the photometric distortions category of transformations. These classes take a function (op) as an input parameter and create objects that apply the op operation to the input image. The op operation is a function that takes an image as input and performs photometric distortions on the input image. These transformations only change the color space, so they do not impact annotations such as bounding boxes or semantic segmentations.

- **The VFlipTransform and HFlipTransform classes** take the original image's height and width and create transformations to flip the input image and its related annotation vertically and horizontally, respectively. The HFlipTransform class is one of the most commonly used transformations.

- **The GridSampleTransform class** takes a flow-field grid with an np.ndarray of size HxWx2. Here, H and W are the height and width of the original input image, and the last dimension of size two indicates the coordinate in the original input image that the output image should sample. The coordinate is normalized in the range of [-1, 1] for the width and height running from [0, W] and [0, H] correspondingly. What this transformation does to the original input depends greatly on this grid. Additionally, by default, no apply_box is implemented for this transformation. Therefore, if required, we can implement a custom transformation class that extends this class and overrides the `apply_box` method.

One example scenario of using this type of transformation is to create barrel distortion or pincushion distortion. *Figure 8.7* illustrates example grids for these types of transformations:

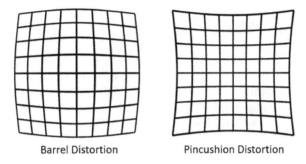

Barrel Distortion   Pincushion Distortion

Figure 8.7: Grids to sample an image and create barrel and pincushion distortions

The following code snippet creates a custom transformation class that uses the cosine wave to create a sample grid and sample from an image. This grid creates a pincushion distortion effect:

```
class PincushionTransform(T.GridSampleTransform):
  def __init__(self, img_h, img_w):
    # initialization codes (explained later)

  def apply_box(self, box: np.ndarray) -> np.ndarray:
    return np.array([self._apply_1_box(b) for b in box])

  def _apply_1_box(self, box):
    # codes to transform one bounding box
```

This class has three methods (`__init__`, `apply_box`, and `_apply_1_box`). The `apply_box` method utilizes the inner method, `_apply_1_box`, to perform transformations on multiple bounding boxes. The `__init__` method consists of the following code:

```
self.img_h = img_h
self.img_w = img_w
# grid
X_range = [np.cos(x) for x in np.linspace(np.pi, 0, 441)]
Y_range = [np.cos(y) for y in np.linspace(np.pi, 0, 543)]
X_range = np.array(X_range)
```

```
Y_range = np.array(Y_range)
X, Y = np.meshgrid(X_range, Y_range)
grid = np.expand_dims(np.stack([X, Y], axis=2), axis=0)
grid = grid.astype('float32')
super().__init__(grid=grid, interp="nearest")
self.X_range = X_range
self.Y_range = Y_range
```

This method (`__init__`) generates a range of cosine values for angles equally distributed from `pi` to 0, with 441 points for the width and 543 points for the height. They are then used to create a mesh grid and initialize an object of its superclass (`GridSampleTransform`). Also, note that the cosine wave in the range of `[np.pi, 0]` returns a range of `[-1, 1]`, as expected by the input parameter for the `GridSampleTransform` class.

`GridSampleTransform` helps transform the input image. However, it does not support the `apply_coords` method. Therefore, we need to code the `apply_box` method to perform box transformations. This `apply_box` method utilizes `_apply_1_box` to perform transformations on all its input boxes. The `_apply_1_box` method is an internal helper method that performs the transformation on a single bounding box. It consists of the following code:

```
gs_box = np.array(box)
img_width = self.img_w
img_height = self.img_h
X_range = self.X_range
Y_range = self.Y_range
x0x1 = 2*gs_box[[0, 2]]/img_width - 1
y0y1 = 2*gs_box[[1, 3]]/img_height - 1
x0x1 = [X_range.searchsorted(x0x1[0]),
        X_range.searchsorted(x0x1[1])]
y0y1 = [Y_range.searchsorted(y0y1[0]),
        Y_range.searchsorted(y0y1[1])]
gs_box = [min(x0x1), min(y0y1), max(x0x1), max(y0y1)]
return gs_box
```

This method (`_apply_1_box`) finds the X and Y coordinates of the top left (x0, y0) and bottom right (x1, y1), normalized by the image's width and height, with the center of the coordinate placed at the center of the image. Normalization ensures that the top-left corner of the image has a [-1, -1] coordinate, the bottom-right corner of the image has a [1, 1] coordinate, and that the origin [0, 0] is at the center of the image. This normalization reflects the image coordinate values after they've been translated into cosine values. Finally, it uses the `searchsorted` method to find the pixel coordinates given the cosine coordinates.

Now, we can use this class in the same way as the other classes that we have learned about so far:

```
fc_tfm = PincushionTransform(*img.shape[:2])
fc_img = fc_tfm.apply_image(img)
fc_boxes = fc_tfm.apply_box(boxes)
visualize_image_boxes(fc_img, boxes = fc_boxes)
```

Figure 8.8 displays the output of this distortion method (right) versus its original input (left):

Original image and ground-truth box   Pincushion image and ground-truth box

Figure 8.8: Original image and the ground-truth box and their pincushion versions

As we can see, this transformation method is important for training models that are robust to the perspective differences due to camera lens artifacts.

- **The CropTransform class** takes the input as the starting coordinate (x0, y0) and the width and height (w and h) and crops it accordingly (input [y0:y0+h, x0:x0+w]). Additionally, you can specify the original width and height (orig_w and orig_h) if you want to make this transformation invertible.

- **The BlendTransform class** takes a source image (src_image), the blending weight of the source image (src_weight), and the blending weight of the destination image (dst_weight). When calling the apply_image(dst_img) method, this transformation blends two images as result = src_image*src_weight + dst_image*dst_weight. Note that the source and destination images must be the same size. Additionally, by default, this transformation does not implement apply_coords or apply_segmentation, and its inverse() method returns the NoOpTransform object. Therefore, it is up to the users to define related methods to decide on how to blend the ground-truth annotations. This class is used to implement the MixUp augmentation technique discussed previously. The following code snippet illustrates the usage of this transformation:

```
src_img = img
dst_img = cv2.imread("Chapter08/4.jpg")
src_weight = 0.7
dst_weight = 0.3
# crop
src_w = src_img.shape[1]
src_h = src_img.shape[0]
cr_tfm = T.CropTransform(0, 0, src_w, src_h)
dst_img = cr_tfm.apply_image(dst_img)
bl_tfm = T.BlendTransform(src_img, src_weight, dst_weight)
bl_img = bl_tfm.apply_image(dst_img)
imshow(bl_img)
```

First, this snippet uses CropTransform to crop the destination image and takes an area the same size as the source image. Then, it uses BlendTransform to mix these two images with ratios of 0.7 for the source and 0.3 for the destination. *Figure 8.9* shows the source and destination images and the output of this blend transformation:

Source Image (*src_img*)   Destination Image (*dst_img*)   Blended Version (*0.7 src_img + 0.3 dst_img*)

Figure 8.9: Source image, destination image, and their blended versions

- **The PadTransform class** pads x0, y0, x1, and y1 pixels with pad_value (the default is 0.0) and seg_pad_value (the default is 0) to the left, top, right, and bottom of the input image and the segmentation mask (if there is one), respectively. Additionally, you can specify the original width and height (orig_w and orig_h) if you want to make this transformation invertible.

- **The TransformList class** extends the Transform class and, thus, has similar methods for performing transformation operations. It accepts the input parameter as a list of transformations (transforms) and maintains it as an attribute. Its method applies all the transformations in this list to the input sequentially. For instance, a TransformList object has a CropTransform object and a PadTransform object in its transforms attribute. Calling the apply_image method of this object first crops the image, then pads it. This scenario helps train robust models to detect objects outside their specific contexts. For example, cropping the snowy background out from the Husky forces the models to learn features from the Husky instead of its background context.

The following code snippet illustrates how to use the CropTransform, PadTransform, and TransformList classes:

```
x0, y0, w, h = (130, 130, 250, 310)
y1, x1 = img.shape[:2] - np.array([y0+h, x0+w])
cr_tfm = T.CropTransform(x0, y0, w, h)
pd_tfm = T.PadTransform(x0, y0, x1, y1)
lst_tfm = T.TransformList([cr_tfm, pd_tfm])
lst_img = lst_tfm.apply_image(img)
lst_boxes = lst_tfm.apply_box(boxes)
visualize_image_boxes(lst_img, boxes = lst_boxes)
```

First, this snippet defines the cropping area (the starting point and width and height), then calculates the required padding pixels for the left, top, right, and bottom of the cropped image to maintain the image size. Then, it creates a list of two corresponding transformations and passes it as the input parameter for the `TransformList` class. Note that calling the `apply_image` and `apply_box` methods of the `TransformList` object executes the two transformations in this list in sequence. *Figure 8.10* shows the output of this snippet (right) versus its original input (left):

Original image and ground-truth box   Cropped and padded version

Figure 8.10: Original image and the ground-truth box and their pincushion versions

Congratulations! By now, you should have mastered the classes Detectron2 provides to help you perform the list of augmentations you learned about in the previous section. However, these classes perform transformations deterministically. In practice, we need to devise some policies with randomness so that we know how to perform these transformations. This is where augmentation classes come in handy.

Augmentation classes

An `Augmentation` class defines policies with randomness to generate a transformation (an object of the `Transform` class) from input data. In other words, it does not perform the transformation but generates a transformation using a given policy. It has a method called `get_transform` that implements the augmentation policy and returns an object of the `Transform` class (or its subclass). The returned `Transform` object then performs transformations in a deterministic way. Therefore, if you want another random transformation, it is necessary to call to `get_transform` again (it does not have to create a new augmentation object) to execute the augmentation policy and generate another transformation.

Detectron2 provides classes to perform commonly used augmentations, and users can define a new class that implements custom augmentations. The following are the common augmentation classes:

The Augmentation class is the base class for all augmentation classes to extend. There are two important methods to note: `get_transform(self, *args)` and `__call__(self, aug_input)`. All its subclasses override the `get_transform` method to accept a list of named input parameters and generate a transformation object. This list of parameters should have names acceptable by `AugInput` (which we'll learn about in the next section). The `__call__` method enables its subclasses to work with `AugInput` directly (more on this in the next section).

The FixedSizeCrop class takes a `crop_size` as a tuple (`height, width`) to crop for that region from the original input randomly. If this size is larger than the original image, it checks whether the `pad` option is `True`. If it is, it pads the input image and segmentation with `pad_value` and `seg_pad_value`, respectively, to the right and bottom (the original image is always in the top-left corner). This class helps produce inputs of the same size, which deep learning models often require. The following code snippet illustrates the use of this augmentation class:

```
fsc_aug = T.FixedSizeCrop(crop_size = (300, 300), pad=True,
pad_value=128)
fsc_tfm = fsc_aug.get_transform(img)
fsc_img = fsc_tfm.apply_image(img)
fsc_boxes = fsc_tfm.apply_box(boxes)
visualize_image_boxes(fsc_img, fsc_boxes)
```

First, this snippet initiates the crop size and sets the padding options, then gets a transformation object using the `get_transform` method. This object can then be used as a deterministic transformation object, as discussed in the previous section. In other words, it crops the same region from different inputs. Additionally, applying this same transformation to inputs with different sizes will not work properly:

Original image and ground-truth box   Cropped versions

Figure 8.11: Original image and the ground-truth box and their cropped versions

The RandomApply class takes the input parameters as an instance of the Transform or Augmentation class (tfm_or_aug) and a probability (prob, which has a default as 0.5) of whether or not to perform such a transformation or augmentation. The following code snippet illustrates the use of this class:

```
ra_aug = T.RandomApply(fsc_tfm, prob = 0.3)
ra_fsc_tfm = ra_aug.get_transform()
ra_fsc_img = ra_fsc_tfm.apply_image(img)
ra_fsc_boxes = ra_fsc_tfm.apply_box(boxes)
visualize_image_boxes(ra_fsc_img, ra_fsc_boxes)
```

If we run this whole snippet several times (getting different Transform objects at other runs), then some of them might be cropped (as in the right of *Figure 8.11*), while some might not (kept the same input as in the left of *Figure 8.11*).

The RandomCrop class takes crop_type and crop_size as input parameters. Let's look at the different crop_type values and how they work:

- "relative": crop_size must be in the range of (0, 1] (normalized by the image size, HxW). Then, it crops a random region in the image whose size is H*crop_size[0], W*crop_size[1].

- "relative_range": crop_size[0] is assigned with a uniformly random sampled value from [crop_size[0], 1]; the same is done for crop_size[1]. The new crop_size is then used as crop_type = "relative".

- "absolute": crop_size must be smaller than the input image's size. It crops a random region in the image whose size is crop_size[0], crop_size[1].

- "absolute_range": crop_size[0] is assigned with a uniformly random sampled value from its current value and the image height ([crop_size[0], H]); the same is done for crop_size[1]. The new crop_size is then used as crop_type = "absolute".

The following code snippet illustrates the usage of this class:

```
rc_aug = T.RandomCrop(
    crop_type="relative",
    crop_size=(0.5, 0.5))
rc_tfm = rc_aug.get_transform(img)
rc_img = rc_tfm.apply_image(img)
rc_boxes = rc_tfm.apply_box(boxes)
visualize_image_boxes(rc_img, rc_boxes)
```

The RandomExtent class takes input parameters as tuples of size two for `scale_range` and `shift_range`. The former helps decide the size of the cropping subregion, while the latter helps determine the location to extract the subregion from. Specifically, the `scale_range` tuple specifies a range with low (l) and high (h) values, where a value (r) is uniformly sampled from this range; then, the output image's size is `r*H` and `r*W` (where H and W are the height and width of the input image, respectively). The second parameter, `shift_range`, is another tuple of size two, `(x, y)`. Two random values (`x_shift` and `y_shift`) are uniformly sampled from the `[-x, x]` and `[-y, y]` ranges; then, the cropped region is also shifted by `x_shift*W/2` and `y_shift*H/2` around the center of the input image. The following code snippet illustrates the usage pattern of this class:

```
re_aug = T.RandomExtent(
    scale_range=(0.5, 0.9),
    shift_range=(0.3, 0.3))
re_tfm = re_aug.get_transform(img)
re_img = re_tfm.apply_image(img)
re_boxes = re_tfm.apply_box(boxes)
visualize_image_boxes(re_img, re_boxes)
```

Figure 8.12 shows some sample output for this code snippet (right) and the original input (left). As we can see, in the original input (left), a dashed box has been placed in the center of the input image. The `shift_range` parameter dictates this box size. A random point within this box is selected to place the center of the cropping region. The `scale_range` parameter decides the size of the cropping region (solid box):

Input and meanings of
scale_range and shift_range

One sample output

Figure 8.12: Original image and the ground-truth box and their random-extent versions

The **RandomFlip class** takes a probability (`prob`) and two Boolean values (`horizontal` and `vertical`). The `prob` parameter is used to decide whether or not to perform flipping; if the outcome is to flip, the other two decide to flip horizontally and vertically. The following code snippet illustrates the usage of this class:

```
rf_aug = T.RandomFlip(prob        = 0.5,
                      horizontal  = True,
                      vertical    = False)
rf_tfm = rf_aug.get_transform(img)
rf_img = rf_tfm.apply_image(img)
rf_boxes = rf_tfm.apply_box(boxes)
visualize_image_boxes(rf_img, rf_boxes)
```

When running this code snippet, as a whole, several times, about 50% of the time, the input flips horizontally, and the input never flips vertically.

The **RandomRotation class** takes `angle`, `expand`, `center`, `sample_style`, and `interp` as its input parameters. The `expand` parameter decides if it should expand the output size so that it fits the rotated image size. The `sample_style` parameter can either be `"range"` or `"choice"`.

If it is `"range"` (default), the following occurs:

- The `angle` parameter is in a range of `[min, max]` values in degrees to uniformly sample one value for rotation
- The `center` parameter is a `[[minx, miny], [maxx, maxy]]` relative interval from which to sample the center, with `[0, 0]` being the top left of the image and `[1, 1]` being the bottom right

If it is `"choice"`, the following occurs:

- The `angle` parameter is a list of values to choose from for rotation
- The `center` parameter is `None`, meaning the center of the rotation is that of the input image

The following code snippet illustrates the usage pattern of this class:

```
rr_aug = T.RandomRotation(
    angle         = [-30, 30],
    sample_style  = "range",
    center        = [[0.4, 0.6], [0.4, 0.6]],
    expand        = False
    )
rr_tfm = rr_aug.get_transform(img)
```

```
rr_img = rr_tfm.apply_image(img)
rr_boxes = rr_tfm.apply_box(boxes)
visualize_image_boxes(rr_img, rr_boxes)
```

This snippet randomly samples a value between -30 and 30 degrees and another value around the center to rotate the input data

The Resize class is pretty straightforward. It takes the target size (`shape=(height, width)`) and the interpolation method (`interp`) as input parameters. Its `get_transform(image)` method returns a `ResizeTransform` object to resize the input image and related annotations into the target size. The following code snippet illustrates the usage of this class:

```
rs_aug = T.Resize(
    shape=(640, 640)
)
rs_tfm = rs_aug.get_transform(img)
rs_img = rs_tfm.apply_image(img)
rs_boxes = rs_tfm.apply_box(boxes)
visualize_image_boxes(rs_img, rs_boxes)
```

This code snippet resizes the input image into a fixed target size. This is a commonly used image preprocessing step because neural networks often require inputs of the same size. However, this technique does not retain the aspect ratios of the inputs if the target has a different ratio.

The RandomResize class takes `shape_list` and `interp` as input parameters. It resizes the input to one item, `(h, w)`, sampled from `shape_list` using the given interpolation method (`interp`). The following code snippet illustrates the usage pattern of this class:

```
rrs_aug = T.RandomResize(
    shape_list=[(320, 320), (640, 640)]
)
rrs_tfm = rrs_aug.get_transform(img)
rrs_img = rrs_tfm.apply_image(img)
rrs_boxes = rrs_tfm.apply_box(boxes)
visualize_image_boxes(rrs_img, rrs_boxes)
```

The ResizeScale class takes `min_scale`, `max_scale`, `target_height`, `target_width`, and `interp` as input parameters. It samples a value in the range of `[min_scale, max_scale]`, scales the target size, then resizes the input so that it fits inside this scaled target size and keeps the original image aspect ratio. The following code snippet illustrates the usage pattern of this class:

```
rss_aug = T.ResizeScale(
    min_scale     = 0.5,
    max_scale     = 1.0,
    target_height = 640,
    target_width  = 640
    )
rss_tfm = rss_aug.get_transform(img)
rss_img = rss_tfm.apply_image(img)
rss_boxes = rss_tfm.apply_box(boxes)
visualize_image_boxes(rss_img, rss_boxes)
```

This code snippet ensures that the scaled image retains the original image ratio and always fits the `640x640` box. This snippet is helpful because we can easily pad the outputs of this snippet to create data that's the same size before feeding it to neural networks.

The ResizeShortestEdge class takes `short_edge_length`, `max_size`, and `sample_style` as input parameters. The output of this class has the same aspect ratio as its input. It resizes the shorter side to a length defined in `short_edge_length`. If the other side of the output is longer than `max_size`, it continues to resize the image to ensure that this longer size is `max_size`. If `sample_style` is `"range"`, then `short_edge_length` is specified as a `[min, max]`, in which a random value is to be sampled. If it is `"choice"`, then `short_edge_length` contains a list of lengths to choose from. This class is commonly used. Detectron2 uses this and RandomFlip as default augmentation techniques for its data augmentation. The following code snippet illustrates the usage pattern of this class:

```
rse_aug = T.ResizeShortestEdge(
    short_edge_length=(640, 672, 704, 736, 768, 800),
    max_size=1333,
    sample_style='choice')
rse_tfm = rse_aug.get_transform(img)
rse_img = rse_tfm.apply_image(img)
rse_boxes = rse_tfm.apply_box(boxes)
visualize_image_boxes(rse_img, rse_boxes)
```

This code snippet randomly selects a value from the specified list of lengths and then resizes the shorter side of the input to this sampled length. If the longer side of the output is larger than 1,333, it continues to resize the image (while keeping the aspect ratio) to ensure that it does not exceed this number.

The RandomCrop and CategoryAreaConstraint classes take `crop_type`, `crop_size`, `single_category_max_area`, and `ignored_category` as input parameters. The first two parameters are similar to those in the `RandomCrop` class. However, this class searches for a cropping area in which no single category occupies a ratio larger than the value specified for `single_category_max_area` for semantic segmentation. This value is specified in 0 to 1 as ratios (set to 1.0 to disable this feature). The last parameter sets the category's `id`, which does not have to follow this constraint. Note that this function applies a best-effort strategy. It only tries, at most, 10 times to achieve the specified conditions. The following code snippet illustrates the usage pattern of this class:

```
cac_aug = T.RandomCrop_CategoryAreaConstraint(
    crop_type                 = "relative",
    crop_size                 = (0.5, 0.5),
    single_category_max_area  = 0.5,
    ignored_category          = None
    )
sem_seg = np.zeros(img.shape[:2])
cac_tfm = cac_aug.get_transform(img, sem_seg=sem_seg)
cac_img = cac_tfm.apply_image(img)
cac_boxes = cac_tfm.apply_box(boxes)
visualize_image_boxes(cac_img, cac_boxes)
```

Note that besides the input image, the `sem_seg` parameter of the `get_transform` method is required because this class is used for semantic segmentation. Thus, this code snippet creates a fake semantic segmentation mask that only contains all zeros for illustration purposes.

The MinIoURandomCrop class takes `min_ious`, `boxes`, `min_crop_size`, `mode_trials`, and `crop_trials` as input parameters. This augmentation attempts to perform a random crop with output that meets the IoU requirement for the input image (controlled by `min_crop_size`) and bounding boxes (controlled by `min_ious`). The `min_crop_size` parameter is in the ratio (ranging from 0 to 1) to the input height and width (`H`, `W`). This class works as follows for the maximum `mode_trials` times, where the cropping region is (`top, left, new_h, new_w`):

I.   Sample for a `min_iou` value from `[0, *min_ious, 1]`.

II.   For the maximum `crop_trials` times, do the following:

    i.   Sample two values from (`min_crop_size*W, W`) and (`min_crop_size*H, H`) to generate `new_h` and `new_w`. It keeps resampling if the crop size ratio is outside the range [0.5, 2.0] or the cropping region becomes a line or point so that it doesn't distort the input too much.

    ii.   Once there's a new size, it samples the `top` and `left` values from [`H - new_h`] and [`W - new_w`]. If the cropping region has an IoU with any ground-truth boxes smaller than `min_iou`, continue to *Step II*.

The following code snippet illustrates the usage pattern of this class:

```
iou_aug = T.MinIoURandomCrop(
    min_ious       = [0.5, 0.6, 0.7],
    min_crop_size = 0.5,
    mode_trials   = 1000,
    crop_trials   = 50
    )
iou_tfm = iou_aug.get_transform(img, boxes=boxes)
iou_img = iou_tfm.apply_image(img)
iou_boxes = iou_tfm.apply_box(boxes)
visualize_image_boxes(iou_img, iou_boxes)
```

Note that besides the input image, the `get_transform` method requires the ground-truth boxes since this augmentation works with bounding boxes. This augmentation is important because it keeps the ground-truth boxes while trying to remove parts of the external contexts. This technique enables training models that focus on learning features from the objects instead of their contexts.

The RandomBrightness class takes `intensity_min` and `intensity_max` as input parameters. It samples a random value (w) from this range and changes the input's pixel values using `output = w*input`. It increases, reserves, or decreases the input intensity if the sampled value is greater than, equal to, or smaller than 1.0, respectively. This augmentation is a photometric distortion; therefore, no change is made to the input annotations. The following code snippet illustrates the usage pattern of this augmentation class:

```
rb_aug = T.RandomBrightness(
    intensity_min = 0.5,
    intensity_max = 1.5
    )
rb_tfm = rb_aug.get_transform(img)
rb_img = rb_tfm.apply_image(img)
visualize_image_boxes(rb_img, boxes)
```

The RandomContrast class works the same way as `RandomBrightness`, except the output is computed as `output = input.mean()*(1-w) + w*input`. The usage pattern is the same as `RandomBrightness`. So, please refer to the previous code snippet as an example.

The RandomSaturation class works the same way as `RandomBrightness`, except it works with the RGB input image format (instead of BGR), and the output is computed as `output = grayscale*(1-w) + w*input`. The grayscale image is computed by the weighted values of the three channels as `[0.299, 0.587, 0.114]` in RGB order (that's why it expects the inputs to be in RGB format). The usage pattern is the same as `RandomBrightness`. So, please refer to the previous code snippet as an example.

The RandomLighting class takes `scale` as an input parameter. This technique uses three eigenvectors (`eigen_vecs`) that correspond to the highest eigenvalues (`eigen_vals`) computed from ImageNet (you can set them to another set of values). The `scale` parameter is the standard deviation of a normal distribution with a mean of zero; this normal distribution helps generate three weights for the three image channels. The eigenvalues are then scaled by these weights accordingly and generate `scaled_eigenvalues`. This class generates a vector called `vec=eigen_vecs.dot(scaled_evenvalues)`. This generated vector is then added to every input image pixel to create the output. This augmentation is a photometric distortion and does not change input annotations. The following code snippet illustrates the usage pattern of this class:

```
rl_aug = T.RandomLighting(scale = 255)
rl_tfm = rl_aug.get_transform(img[:,:,::-1])
rl_img = rl_tfm.apply_image(img[:, :, ::-1])
visualize_image_boxes(rl_img[:, :, ::-1], boxes)
imshow(img-rl_img[:,:,::-1])
```

Note that we may need to normalize the input image or choose a high value for `scale`. The default eigenvectors and values are for RGB images, while this snippet works with BGR images. Therefore, this snippet keeps reversing the color channels back and forth.

The AugmentationList class takes a list of transformations or augmentations (`augs`) as an input parameter. This class helps apply changes (`augs`) to the input sequentially. This class does not implement the `get_transform()` method; instead, it utilizes the `__call__` method to apply the augmentations or transformations to the input wrapped inside `AugInput`.

The following section provides further details about `AugInput` and its usage pattern.

The AugInput class

This special class encapsulates data for inputs to augmentation objects. Specifically, an instance of this class includes attributes for the input image (`image`) and related annotations, including bounding boxes (`boxes`) and semantic segmentations (`sem_seg`). This class has a method called `transform(tfm: Transform)` that takes the input parameter as a `Transform` object (`tfm`) and performs the corresponding transformation on all of its attributes (`image`, `boxes`, and `sem_seg`) in place. If required, users can define a similar class with additional attributes to be transformed by providing attribute access (`getattr`) and a corresponding `transform` method to perform additional transformations on the attributes in place and return a `Transform` object.

Despite having the `transform` method, which can be used by itself, the main usage pattern of this class is to be used as the input to the `__call__` method of an `Augmentation` object. In this case, the `transform` operation of the `Augmentation` class changes the input in place and returns an object of a `Transform` class to perform augmentation on other inputs if needed. This method utilizes the attribute names from `aug_input` (an object of the `AugInput` class), passes the corresponding attribute values to the `get_transform` method, generates a transformation object, and then performs the corresponding transformation on the data in the `aug_input` object. One sample usage of the `AugInput` class can be seen in the following code snippet:

```
# AugmentationList
rse_aug = T.ResizeShortestEdge(
    short_edge_length = (640, 672, 704, 736, 768, 800),
    max_size          = 800,
    sample_style      = 'choice')
rf_aug = T.RandomFlip(
    prob         = 0.5,
    horizontal  = True,
    vertical    = False)
augs = [rse_aug, rf_aug]
augmentation = T.AugmentationList(augs)
```

```
# AugInput
input = T.AugInput(img, boxes=boxes)
# Transformation
tfms = augmentation(input)
tfm_img = input.image
tfm_boxes = input.boxes
visualize_image_boxes(tfm_img, tfm_boxes)
```

This snippet shows the most commonly used augmentation in Detectron2. This snippet creates a list of two augmentations for resizing and randomly flipping the input horizontally. In this case, the augmentation is performed on the input in place. Thus, input.image and input.boxes return the transformed versions of these attributes. Furthermore, the returned object (tfms) is an object of the Transform object in the augmentation object that can be used to change other inputs. This snippet shows the exact default augmentations used in Detectron2. However, it helps to illustrate that it is relatively easy to plug in existing image augmentations and transformations using the declarative approach. If you need to create a custom image augmentation technique, then read on – the next chapter will show you how to create custom image augmentations, as well as how to apply image augmentations at test time.

Congratulations! At this point, you should have mastered important image augmentation techniques and have profound knowledge of the transformations and augmentations that Detectron2 offers.

Summary

This chapter introduced image augmentations and why it is essential to perform them in computer vision. Then, we covered common and state-of-the-art image augmentation techniques. After understanding the theoretical foundation, we looked at Detectron2's image augmentation system, which has three main components, and their related classes: Transform, Augmentation, and AugInput. Detectron2 provides a declarative approach for applying existing augmentations conveniently.

The existing system supports augmentations on a single input, while several modern image augmentations require data from different inputs. Therefore, the next chapter will show you how to modify several Detectron2 data loader components so that you can apply modern image augmentation techniques. The next chapter also describes how to apply test time augmentations.

Applying Train-Time and Test-Time Image Augmentations

The previous chapter introduced the existing augmentation and transformation classes Detectron2 offers. This chapter introduces the steps to apply these existing classes to training. Additionally, Detectron2 offers many image augmentation classes. However, they all work on annotations from a single input at a time, while modern techniques may need to combine annotations from different inputs while creating custom augmentations. Therefore, this chapter also provides the foundation for Detectron2's data loader component. Understanding this component helps explain how to apply existing image augmentations and modify existing codes to implement custom techniques that need to load data from different inputs. Finally, this chapter details the steps for applying image augmentations during test time to improve accuracy.

By the end of this chapter, you will be able to understand how Detectron2 loads its data, how to apply existing image augmentation classes, and how to implement custom image augmentation techniques that need to load data from multiple inputs. Additionally, you can also use image augmentation techniques during inference time. Specifically, this chapter covers the following topics:

- The Detectron2 data loader
- Applying existing image augmentation techniques
- Developing custom image augmentation techniques
- Applying test-time image augmentation techniques

Technical requirements

You should have completed *Chapter 1* to have an appropriate development environment for Detectron2. All the code, datasets, and results are available on the GitHub repo of the book at `https://github.com/PacktPublishing/Hands-On-Computer-Vision-with-Detectron2`.

The Detectron2 data loader

Applying augmentations in Detectron2 can be straightforward and complicated at the same time. It is relatively easy to use the declarative approach and apply existing transformations and augmentations provided by Detectron2, which should meet the most common needs. However, adding custom augmentations that require multiple data samples (e.g., MixUp and Mosaic) is a little complicated. This section describes how Detectron2 loads data and how to incorporate existing and custom data augmentations into training Detectron2 models. *Figure 9.1* illustrates the steps and main components of the Detectron2 data loading system.

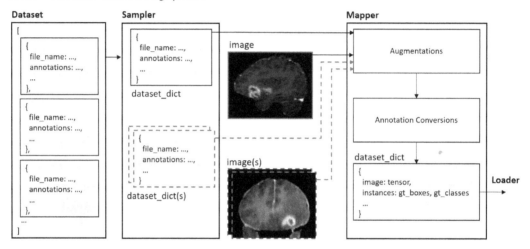

Figure 9.1: Loading data and data augmentations in Detectron2

There are classes for `Dataset`, `Sampler`, `Mapper`, and `Loader`. The `Dataset` component normally stores a list of data items in JSON format. The `Sampler` component helps to randomly select one data item (`dataset_dict`) from the dataset. The selected data item has texts for the image file name and annotations. Thus, the `Mapper` loads the actual image. After that, the annotations and loaded image go through augmentations and data conversion into an image tensor and instances for annotations. The `Mapper` produces one data item, and the `Loader` helps to pack these in batches and feed them to training. A few augmentation techniques may want to load extra samples (`dataset_dict(s)` and `image(s)`) besides the current input, and these items are expressed in dashed lines in *Figure 9.1*. Once you understand how Detectron2 loads its data, it is relatively easy to incorporate existing data augmentations into Detectron2 training.

Applying existing image augmentation techniques

Augmentations are in the `Mapper` component. A `Mapper` receives a list of augmentations and applies them to the image and annotations accordingly. The following code snippet creates a Detectron2 trainer and specifies a list of existing augmentations to use:

```
class MyTrainer(DefaultTrainer):
  @classmethod
  def build_train_loader(cls, cfg):
    augs = []
    # Aug 1: Add RandomBrightness with 50% chance
    # Aug 2: Add ResizeShortestEdge
    # Aug 3: Add RandomFlipping
    mapper = DatasetMapper(cfg,
                           is_train        = True,
                           augmentations   = augs)
    return build_detection_train_loader(cfg, mapper=mapper)
```

Specifically, this snippet creates a new trainer class (`MyTrainer`) that overrides the `build_train_loader` class method. This method declares a list of augmentations, including performing `RandomBrightness` 50% of the times and applying `ShortestEdgeResize` and `RandomFlip` with parameters loaded from the configuration file. This method then creates a `Mapper` instance that uses this list of augmentations. Finally, the `build_detection_train_loader` built-in method utilizes this `Mapper` object to generate a `Loader` object to load data for training. This class (`MyTrainer`) can then be used to train models in the same way we use the `DefaultTrainer` class.

The code to add a `RandomBrightness` augmentation object with a 50% chance of being applied (`RandomApply`) is as follows:

```
augs.append(
    T.RandomApply(
        T.RandomBrightness(
            intensity_min = 0.5,
            intensity_max = 1.5),
        prob = 0.5
        )
    )
```

The second augmentation type used is `ResizeShortestEdge`, and the following code snippet creates and appends that augmentation:

```
min_size = cfg.INPUT.MIN_SIZE_TRAIN
max_size = cfg.INPUT.MAX_SIZE_TRAIN
sample_style = cfg.INPUT.MIN_SIZE_TRAIN_SAMPLING
augs.append(
    T.ResizeShortestEdge(
        min_size,
        max_size,
        sample_style)
    )
```

Lastly, the following code snippet adds a `RandomFlipping` augmentation to the list of augmentations used by this custom trainer:

```
if cfg.INPUT.RANDOM_FLIP != "none":
  augs.append(
    T.RandomFlip(
      horizontal=cfg.INPUT.RANDOM_FLIP == "horizontal",
      vertical=cfg.INPUT.RANDOM_FLIP == "vertical",
    )
  )
```

It is relatively easy to use this declarative approach and incorporate different augmentations/transformations using classes provided by Detectron2, as shown in the `MyTrainer` class. These existing classes only work with one data item (one image and its related annotations). However, several useful image augmentation techniques may require data from multiple samples. For instance, MixUp loads inputs from two data samples, and Mosaic loads inputs from four. In these cases, it is necessary to rewrite several components of the Detectron2 data loader system to enable combining annotations from different samples into one. The next section illustrates these scenarios.

Developing custom image augmentation techniques

Suppose the custom augmentation requires annotations loaded from one data sample (`dataset_dict`). In that case, it is relatively simple to implement the custom augmentation or transformation and incorporate it using the declarative approach described in the previous section. This section focuses on more complicated augmentation types that require loading and combining inputs from multiple samples. In these cases, we need to rewrite several parts of the Detectron2 data loader system. Thus, this section describes modifications to the Detectron2 data loader system and develops two custom image augmentations (MixUp and Mosaic) for illustration purposes.

Modifying the existing data loader

The following code snippet imports some required packages and creates an extended version (ExtendedAugInput) of the AugInput class that also enables passing a dataset_dict parameter. The reason is that augmentations such as MixUp and Mosaic may add annotations from other input samples into the current input sample. These additional annotations need to be added to the current dataset_dict:

```
# some import statements are removed for space efficiency
class ExtendedAugInput(T.AugInput):
  def __init__(self, image, *,
                  boxes        = None,
                  sem_seg      = None,
                  dataset_dict = None):
    super().__init__(image, sem_seg=sem_seg, boxes=boxes)
    self.dataset_dict = dataset_dict
    self.is_train = True
```

Next, we also need to override DatasetMapper to create an extended Mapper class (ExtendedDatasetMapper):

```
class ExtendedDatasetMapper(DatasetMapper):
  def _convert_annotations(self,
                            dataset_dict,
                            boxes,
                            image_shape):
    # Codes are explained later
  def __call__(self, dataset_dict):
    # Step 1: Read images and annotations from dataset_dict
    # Step 2: Perform transformations/augmentations
    # Step 3: Convert annotations to instances
    return dataset_dict
```

This extended Mapper does two modifications to the standard one. First, it uses ExtendedAugInput as the input data type to pass dataset_dict to the augmentation objects and allows them to modify dataset_dict. Second, it creates a method called _convert_annotations to convert dataset_dict to the format expected by the Loader component (with image tensor and annotation instances). This method also considers the case that annotations were added to dataset_dict by the augmentations.

The following is the code snippet for the `_convert_to_annotations` method:

```
annos = []
for i, annotation in enumerate(dataset_dict.
pop("annotations")):
  bbox = boxes[i]
  annotation["bbox"] = bbox
  annotation["bbox_mode"] = BoxMode.XYXY_ABS
  annos.append(annotation)
instances = utils.annotations_to_instances(
    annos, image_shape,
    mask_format=self.instance_mask_format
)
if self.recompute_boxes:
  gt_boxes = instances.gt_masks.get_bounding_boxes()
  instances.gt_boxes = gt_boxes
  ft_instances = utils.filter_empty_instances(instances)
  dataset_dict["instances"] = ft_instances
```

This method takes the input as the dataset dictionary (`dataset_dict`) and converts the annotations in `dataset_dict` into instances format using the Detectron2 utility function (`annotations_to_instances`). This method is similar to the `_transform_annotations` method in the `DatasetMapper` class. However, we need to override this so that it takes the transformed bounding boxes instead of the bounding boxes in the original `dataset_dict` before converting annotations to instances.

We need to override the `__call__` method of the `DatasetMapper` class in its extended version because we need to modify the current `dataset_dict` to incorporate more annotations loaded from other sampled inputs. This method has three main steps, which are detailed here:

Step 1: This step reads the image and related annotation data into `dataset_dict` using the information specified in the input `dataset_dict`. This step is exactly the same as that in its original version:

```
dataset_dict = copy.deepcopy(dataset_dict)
image = utils.read_image(dataset_dict["file_name"],
                         format=self.image_format)
utils.check_image_size(dataset_dict, image)
if "sem_seg_file_name" in dataset_dict:
    sem_seg_gt = utils.read_image(
        dataset_dict.pop("sem_seg_file_name"), "L"
```

```
        ).squeeze(2)
else:
    sem_seg_gt = None

boxes = np.array([
    BoxMode.convert(
        obj["bbox"],
        obj["bbox_mode"], BoxMode.XYXY_ABS)
    for obj in dataset_dict['annotations']])
```

Step 2: This step creates an object of `ExtendedAugInput` to include `dataset_dict` as an input parameter because we need to modify this `dataset_dict` to add further bounding boxes from other sampled inputs. Except for the additional modification to `dataset_dict`, all other steps remain the same as in the original version of the `DatasetMapper` class:

```
aug_input = ExtendedAugInput(image,
                    sem_seg       = sem_seg_gt,
                    dataset_dict = dataset_dict,
                    boxes = boxes)

transforms      = self.augmentations(aug_input)
image           = aug_input.image
sem_seg_gt      = aug_input.sem_seg
dataset_dict    = aug_input.dataset_dict
boxes           = aug_input.boxes

image_shape = image.shape[:2]
dataset_dict["image"] = torch.as_tensor(
    np.ascontiguousarray(image.transpose(2, 0, 1)))
```

Step 3: This last step utilized the `_convert_annotations` method described previously to convert transformed/augmented data listed in `dataset_dict` into instances to be processed by Detectron2:

```
if "annotations" in dataset_dict:
  self._convert_annotations(dataset_dict,
                            boxes,
                            image_shape)
```

The following code snippet creates a Sampler component that helps to sample extra inputs while performing augmentations such as MixUp and Mosaic:

```
from detectron2.data import DatasetCatalog
class DataDictSampler():
  def __init__(self, name_ds):
    ds = DatasetCatalog.get(name_ds_train)
    self.ds = ds
  def get_items(self, n=3):
    indices = np.random.randint(
        low   = 0,
        high  = len(self.ds)-1,
        size  = n)
    return [copy.deepcopy(self.ds[_]) for _ in indices]
```

This code snippet takes a dataset name registered with Detectron2 and loads its data (in JSON format). It then builds a method (get_items) to randomly sample some inputs for performing augmentations. A deep copy of the data items is necessary because the augmentations may change the original data.

Congratulations! After having these extended classes, we are ready to write custom augmentation classes that combine annotations from different inputs.

Developing the MixUp image augmentation technique

MixUp is a useful and modern image augmentation technique introduced in the previous chapter. Detectron2 does not provide a built-in class for this image augmentation technique. Therefore, this section provides the steps to develop this image augmentation technique. The following code snippet creates the MixUp augmentation class:

```
class MixUpAug(T.Augmentation):
  def __init__(self, cfg, src_weight=0.5, dst_weight=0.5):
    self.cfg = cfg
    self.sampler = DataDictSampler(cfg.DATASETS.TRAIN[0])
    self.src_weight = src_weight
    self.dst_weight = dst_weight
  def get_transform(self, image, dataset_dict):
    cfg = self.cfg
    # Step 1: get one more random input
    # Step 2: append annotations and get mix-up boxes
```

```
    # Step 3: return an object of MixUpTransform
def __repr__(self):
    sw, dw = self.src_weight, self.dst_weight
    return f"MixUp(src {sw}, dst {dw})"
```

This Augmentation class uses the built sampler object to select another input to mix with the current one. It also adds annotations from this extra image to the current dataset_dict and passes the loaded image and additional bounding boxes to create a new MixUpTransform object (described later). This class extends the Augmentation class and overrides three methods. The __init__ method is straightforward, taking the inputs and assigning them to corresponding data members. Additionally, it also creates an instance of the DataDictSampler class to help sampling one more random input for performing the mixup. The __repr__ method is used to represent the objects of this augmentation type in the string format for messaging/logging purposes.

The get_transform method is the core method of this class and it has three main steps:

Step 1: This step samples an extra input from the input dataset to perform a mixup with the current input using an object of the DataDictSampler class:

```
ds_dict = self.sampler.get_items(n=1)[0]
mu_image = utils.read_image(ds_dict["file_name"],
                            format=cfg.INPUT.FORMAT)
utils.check_image_size(ds_dict, mu_image)
```

Step 2: This step takes the annotations from the extra (mixup) image and adds them to the list of annotations in the dataset_dict of the current input. It also extracts the bounding boxes from this extra input and stores them into mu_boxes (mixup boxes) to be used in the MixUpTransform class:

```
annotations = ds_dict["annotations"]
dataset_dict["annotations"] += annotations
mu_boxes = np.array([
    BoxMode.convert(
        obj["bbox"], obj["bbox_mode"],
        BoxMode.XYXY_ABS)
    for obj in annotations
    ])
```

Step 3: This step takes all the generated data and passes it as inputs for creating a MixUpTransform object to actually perform the transformation:

```
return MixUpTransform(image       = image,
                      mu_image     = mu_image,
```

```
                              mu_boxes      = mu_boxes,
                              src_weight  = self.src_weight,
                              dst_weight  = self.dst_weight)
```

The following code snippet creates the `MixUpTransform` class:

```
class MixUpTransform(T.Transform):
    def __init__(self, image, mu_image, mu_boxes,
                 src_weight = 0.5,
                 dst_weight = 0.5):
        # Step 1: resize the mu_image and mu_boxes
        # Step 2: pad mu_image
        # Step 3: save values
    def apply_image(self, image):
        bl_tfm = T.BlendTransform(src_image   =self.mu_image,
                                  src_weight =self.src_weight,
                                  dst_weight =self.dst_weight)
        return bl_tfm.apply_image(image)
    def apply_coords(self, coords):
        return coords
    def apply_box(self, boxes):
        boxes = np.vstack([boxes, self.mu_boxes])
        return boxes
```

This transformation class has three main methods. The __init__ method resizes the extra image and related annotations to the current input size using the `ResizeShortestEdge` class and padding (with NumPy). Its `apply_image` method utilizes the `BlendTransform` class and mixes the two images. Its `apply_coords` method merely returns its input (does nothing), and its `apply_box` method combines the bounding boxes from the two samples.

The __init__ method has the following three steps:

Step 1: This step uses the `ResizeShortestEdge` class to resize the extra image (mixup image) to fit into the current input image:

```
image_size = image.shape[:2]
rse = T.ResizeShortestEdge([min(image_size)],
                            min(image_size),
                            "choice")
aug_i = T.AugInput(image=mu_image, boxes = mu_boxes)
```

```
rse(aug_i)
mu_image, mu_boxes = aug_i.image, aug_i.boxes
```

Step 2: This step creates an image the same size as the input image and places mu_image into it. In other words, this step is similar to padding zeros to mu_image to make it the same size as the current input image:

```
img = np.zeros_like(image).astype('float32')
img[:mu_image.shape[0], :mu_image.shape[1], :]=mu_image
```

Step 3: This last step takes the resized image and bounding boxes and stores them for future use:

```
self.mu_image    = img
self.mu_boxes    = mu_boxes
self.src_weight  = src_weight
self.dst_weight  = dst_weight
```

Figure 9.2 shows an example of the MixUp Augmentation class that combines images and bounding boxes from two inputs:

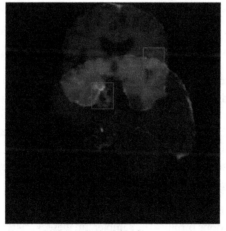

MixUp Example

Figure 9.2: Example of the MixUp Augmentation class

This Augmentation class can then be used in a way similar to other existing augmentation classes in Detectron2 using the declarative approach.

Similar steps can be used to create the MosaicAugmentation class. These steps are covered in detail in the next section.

Developing the Mosaic image augmentation technique

The Mosaic image augmentation technique needs to combine three more inputs to the current input and create an augmented version (further detail was provided in the previous chapter). There are also two classes for this technique, one for the Augmentation class and one for the Transformation class. The following code snippet creates the MosaicAugmentation class:

```
class MosaicAug(T.Augmentation):
  def __init__(self, cfg):
    self.cfg = cfg
    self.sampler = DataDictSampler(cfg.DATASETS.TRAIN[0])
  def get_transform(self, image, dataset_dict):
    cfg = self.cfg
    mo_items   = self.sampler.get_items()
    mo_images  = []
    mo_boxes   = []
    # extract mo_images, mo_boxes and update dataset_dict
    mt = MosaicTransform(mo_images, mo_boxes)
    return mt
  def __repr__(self):
    return "MosaicAug(4)"
```

This class utilizes the data sampler built previously and samples for three more inputs. It then extracts the images and bounding boxes from these extra inputs. It also adds the annotations from the extra inputs to the current list of annotations to dataset_dict so the next step can utilize them to generate annotation instances in the format supported by Detectron2. Specifically, the __init__ method creates a data sampler object (sampler) to sample for three more extra inputs for performing the Mosaic task. The __repr__ method is to give a string representation of this object for messaging/logging purposes.

The core method of this class is the get_transform method. It first samples three more extra images using the data sampler. It then loops through the sampled images, extracts their images and bounding boxes, and stores them into mo_images and mo_boxes, respectively. Additionally, it extracts the annotations from these extra images and adds them into the current dataset_dict so they can be converted into Detectron2 annotation instances respectively. Finally, it passes the extracted data as the inputs to generate an object of the MosaicTransform class to perform the transformation:

```
for ds_dict in mo_items:
  # mo_images
  mo_image = utils.read_image(ds_dict["file_name"],
                              format=cfg.INPUT.FORMAT)
```

```
utils.check_image_size(ds_dict, mo_image)
mo_images.append(mo_image)
annotations = ds_dict["annotations"]
# mo_boxes
mo_boxes.append(np.array([
    BoxMode.convert(
        obj["bbox"],
        obj["bbox_mode"],
        BoxMode.XYXY_ABS)
    for obj in annotations]))
# add annotations
dataset_dict["annotations"] += annotations
```

Combining images and annotations for the Mosaic image augmentation technique is a little involved, and there are many different ways to combine them. *Figure 9.3* illustrates the approach adopted in this section:

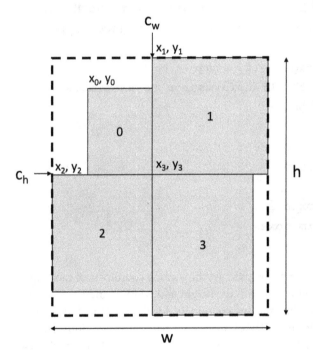

Figure 9.3: Combining images and annotations in Mosaic image augmentation

Specifically, the current input image is indexed as 0 and is combined with three more sampled images (they are indexed as 1, 2, and 3). We first need to find the size (h and w for height and weight) of the combined image (dashed line). The height is the sum of the maximum heights of the top pair (0 and 1) and bottom pair (2 and 3). Similarly, the width of the output image is the sum of the maximum widths of the left pair (0 and 2) and the right pair (1 and 3) of images. Additionally, four images are combined at the four quadrants created by two lines placed at c_h and c_w, correspondingly. It is then relatively easy to find the start point to place each image (x_i, y_i) and how to move their corresponding annotations using the generated numbers (w, h, c_h, and c_w).

The following class provides the code snippet for the MosaicTransformation class:

```
class MosaicTransform(T.Transform):
  def __init__(self, mo_images, mo_boxes):
    self.mo_images   = mo_images
    self.mo_boxes    = mo_boxes

  def get_loc_info(self, image):
    # extract location information for the Mosaic inputs
    return (h, w, ch, cw, widths, heights, x_pads, y_pads)

  def apply_image(self, image):
    # perform the transformation to the image
    return output

  def apply_coords(self, coords):
    return coords

  def apply_box(self, boxes):
    # transform boxes
    return boxes
```

Observably, get_loc_info helps to get the starting coordinate of each input. The aply_image method creates a default image with the height and width found (h and w) and places each image content on its corresponding starting coordinate. Ideally, we should override the apply_coords method to make this technique work with other types of annotations. However, for simplicity, we only override apply_boxes directly by combining boxes and shifting them according to the input start coordinate accordingly.

The `get_loc_info` method implements the steps described in *Figure 9.3* to get the corresponding start locations, the center lines, the widths, and the heights of the four inputs as follows:

```
images = [image] + self.mo_images
heights = [i.shape[0] for i in images]
widths = [i.shape[1] for i in images]
ch = max(heights[0], heights[1])
cw = max(widths[0], widths[2])
h = (max(heights[0], heights[1]) +
     max(heights[2], heights[3]))
w = (max(widths[0], widths[2]) +
     max(widths[1], widths[3]))
# pad or start coordinates
y0, x0 = ch-heights[0], cw - widths[0]
y1, x1 = ch-heights[1], cw
y2, x2 = ch, cw - widths[2]
y3, x3 = ch, cw
x_pads = [x0, x1, x2, x3]
y_pads = [y0, y1, y2, y3]
```

The `apply_image` method creates an output image and places the four input images into their respective locations generated by the `get_loc_info` method:

```
# get the loc info
self.loc_info = self.get_loc_info(image)
h, w, ch, cw, widths, heights, x_pads, y_pads=self.loc_info
output = np.zeros((h, w, 3)).astype('float32')
images = [image] + self.mo_images
for i, img in enumerate(images):
  output[y_pads[i]: y_pads[i] + heights[i],
         x_pads[i]: x_pads[i] + widths[i],:] = img
```

The `apply_box` method combines the bounding boxes from the current input and the three extra inputs. It then moves/pads these bounding boxes with respect to the new locations of each input image:

```
# combine boxes
boxes = [boxes] + self.mo_boxes
# now update location values
_, _, _, _, _, _, x_pads, y_pads = self.loc_info
```

```
for i, bbox in enumerate(boxes):
  bbox += np.array(
            [x_pads[i], y_pads[i], x_pads[i], y_pads[i]]
         )
# flatten it
boxes = np.vstack(boxes)
```

The `MosaicAugmentation` class can then be used in the same way as other existing Detectron2 augmentation classes. *Figure 9.4* shows an example of the output of the Mosaic augmentation applied to the brain tumors dataset:

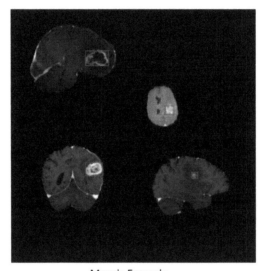

Mosaic Example

Figure 9.4: Example of Mosaic augmentation

This version of the input has four images and their corresponding annotations. This augmentation technique is helpful, especially when the batch size is small, and we would like to use this technique to increase batch sizes. Additionally, it may help to improve the accuracy of predicting annotations with smaller bounding box sizes.

Congratulations! By this time, you should be able to apply existing Detectron2 augmentations or create custom augmentations for training Detectron2 models. As an exercise, you can create a trainer that utilizes these classes and train models for detecting brain tumors. Due to space limitations, this section does not list the source codes for the Trainer that utilizes these augmentations. Please refer to the Jupyter notebooks provided for the source code for this chapter.

This section covered the steps to apply image augmentations to the training process. In many cases, similar techniques can be utilized at test time to improve accuracy, and the following section covers these techniques and shows you how to apply test time augmentations.

Applying test-time image augmentation techniques

Test-time augmentations (**TTA**) can improve prediction performance by providing different versions of the input image for predictions and performing **non-maximum suppression** (**NMS**) on the resulting predictions. Detectron2 provides two classes for this: `DatasetMapperTTA` and `GeneralizedRCNNWithTTA`. The `DatasetMapperTTA` class helps to map a dataset dictionary (a data item in JSON format) into the format expected by Detectron2 models with the opportunity to perform augmentations. The default augmentations used are `ResizeShortestEdge` and `RandomFlip`. The `GeneralizedRCNNWithTTA` class takes the original model and the `Mapper` object as inputs. It performs predictions on the augmented data and preprocesses the resulting outputs.

Let us use the code approach to explain these two classes. As a routine, we first install Detectron2, load the brain tumors dataset, and register the test dataset. Next, the following code snippet gets a pre-trained model:

```
output_path = "output/object_detector_hook/"
!wget --quiet <url_to/Chapter08/object_detector_hook.zip>
!unzip --q object_detector_hook.zip
!unzip --q {output_path}/model_best.pth.zip
```

The following code snippet loads the pre-trained model configuration:

```
# get a configuration
with open(output_path + "cfg.pickle", "rb") as f:
  cfg = pickle.load(f)
```

The following configurations parameters for TTA are available (and settable):

```
# current TTA settings
print("min_sizes", cfg.TEST.AUG.MIN_SIZES)
print("max_size", cfg.TEST.AUG.MAX_SIZE)
print("flip", cfg.TEST.AUG.FLIP)
```

This code snippet should display these default values:

```
min_sizes (400, 500, 600, 700, 800, 900, 1000, 1100, 1200)
max_size 4000
flip True
```

Observably, per one input image, there are ten other augmented versions (nine for resizes and one for flipping).

The following code snippet then gets the testing dataset and creates a DatasetMapperTTA object with the previous augmentations:

```
# device
device = "cuda" if torch.cuda.is_available() else "cpu"
cfg.MODEL.DEVICE = device
# get the dataset
ds = DatasetCatalog.get(name_ds_test)
cfg.MODEL.ROI_HEADS.SCORE_THRESH_TEST = 0.7
cfg.MODEL.WEIGHTS = os.path.join(output_path,
                                 "model_best.pth")
model = DefaultPredictor(cfg).model
```

After getting the model, the following code snippet creates a GeneralizedRCNNWithTTA object and performs predictions with augmentations:

```
# use default DatasetMapperTTA
min_sizes   = cfg.TEST.AUG.MIN_SIZES
max_size    = cfg.TEST.AUG.MAX_SIZE
flip      = cfg.TEST.AUG.FLIP
tta_mapper  = DatasetMapperTTA(min_sizes, max_size, flip)
# the tta model
tta_model   = GeneralizedRCNNWithTTA(
    cfg, model,
    tta_mapper = tta_mapper,
    batch_size = cfg.SOLVER.IMS_PER_BATCH)
with torch.no_grad():
  ret = tta_model(ds)
```

The next code snippet then prepares a helper method that can display the input image in its original size:

```
def imshow(image):
    dpi = plt.rcParams["figure.dpi"]
    im_data = image[:,:, ::-1]
    height, width, depth = im_data.shape
    figsize = width / float(dpi), height / float(dpi)
    fig = plt.figure(figsize=figsize)
    plt.imshow(im_data)
    plt.imshow(im_data)
    plt.axis("off")
    plt.show()
```

This imshow method can then be utilized in the visualize_output method to visualize the predicted outputs to help qualitatively evaluate the results:

```
def visualize_output(img, output, cfg, scale=1.0):
    v = Visualizer(img[:, :, ::-1],
                    MetadataCatalog.get(cfg.DATASETS.TEST[0]),
                    scale=scale)
    annotated_img = v.draw_instance_predictions(
        output["instances"].to("cpu"))
    imshow(annotated_img.get_image())
```

Finally, the following code snippet visualizes some prediction results:

```
import numpy as np
for idx in np.random.randint(0, len(ds), 3):
    img = cv2.imread(ds[idx]['file_name'])
    output = ret[idx]
    visualize_output(img, output, cfg)
```

In many cases, TTA improves prediction accuracy. However, TTA significantly adds to the inference time (about 10-fold in this specific case). Additionally, to quantitatively evaluate whether test time augmentations help improve prediction quality, it would be better to evaluate using F1-score instead.

Congratulations! By this time, you should have mastered important image augmentation techniques, have profound knowledge about Detectron2's image augmentation system, and know how to apply train-time and test-time augmentations.

Summary

This chapter described the steps to apply image augmentation techniques using Detectron2 at both train time and test time (inferencing time). Detectron2 provides a declarative approach to applying existing augmentations conveniently. However, the current system supports augmentations on a single input, while several modern image augmentations require data from different inputs. Therefore, this chapter described the Detectron2 data loader system and provided steps to modify several Detectron2 data loader components to enable applying modern image augmentation techniques such as MixUp and Mosaic that require multiple inputs. Lastly, this chapter also described the features in Detectron2 that allow for performing test-time augmentations.

Congratulations! You now understand the Detectron2 architecture for object detection models and should have mastered the steps to prepare data, train, and fine-tune Detectron2 object detection models. The following part of this book has a similar structure and contents dedicated to Detectron2 instance segmentation models.

Part 3: Developing a Custom Detectron2 Model for Instance Segmentation Tasks

The third part is another hands-on project. This part provides the readers with common tools for collecting and labeling images for object instance segmentation tasks. Additionally, it covers the steps to extract data from different sources and reconstruct a dataset in the format supported by Detectron2. Before training an object segmentation model, this section also utilizes the codes and visualizations approach to explain the architecture of an object segmentation application developed using Detectron2. The object instance segmentation models utilize results from the object detection models. Therefore, all the techniques introduced in the previous part for fine-tuning object detection models work the same for object instance segmentation models. However, object instance segmentation has an important feature to fine-tune: the quality of the boundaries of the detected objects. Therefore, this section also introduces PointRend, a project inside Detectron2 that helps improve the sharpness of the object's boundaries. Specifically, this part provides codes to use existing PointRend models and steps to train custom PointRend models on custom datasets.

The third part covers the following chapters:

- *Chapter 10, Training Instance Segmentation Models*
- *Chapter 11, Fine-Tuning Instance Segmentation Models*

10

Training Instance Segmentation Models

This chapter provides you with common tools for collecting and labeling images for object instance segmentation tasks. Additionally, it covers the steps to extract data from different sources and reconstruct a dataset in the format supported by Detectron2. Before training an object segmentation model, this chapter also utilizes the code and visualizations approach to explain the architecture of an object segmentation application developed using Detectron2.

By the end of this chapter, you will understand different techniques to collect and label data for training models with object instance segmentation tasks. You will have hands-on experience constructing a custom dataset by extracting data from a nonstandard format and reconstructing it in the format supported by Detectron2 for object detection. Additionally, you will have profound knowledge about the architecture and hands-on experience of training an object instance segmentation application in Detectron2 on a custom dataset. Specifically, this chapter covers the following topics:

- Preparing data for training segmentation models
- The architecture of the segmentation models
- Training custom segmentation models

Technical requirements

You should have completed *Chapter 1* to have an appropriate development environment for Detectron2. All the code, datasets, and results are available on the GitHub repo of the book at `https://github.com/PacktPublishing/Hands-On-Computer-Vision-with-Detectron2`.

Preparing data for training segmentation models

This section introduces several scenarios about data preparation steps that may be necessary for your cases. Specifically, this section first introduces the common tasks to prepare a dataset, including getting images, labeling images, and converting annotations. Additionally, in practice, data may come in various formats that might not be standard, and in this case, we may need to perform the data preparation steps from scratch. Therefore, this section also introduces the steps to prepare the brain tumor dataset for training custom segmentation models using Detectron2.

Getting images, labeling images, and converting annotations

If you do not have a dataset, *Chapter 3* introduces common places to obtain data for computer vision applications in general. It would help if you also went to these sources for object instance segmentation data. Also, you can use the same Python script (`Detectron2_Chapter03_Download_Images.ipynb`) introduced in *Chapter 3* to download images from the internet and perform labeling.

Chapter 3 also introduced several online and offline tools for data labeling that support instance segmentation labeling (polygon labeling). Like `labelImg`, commonly used for bounding box labeling (object detection task), `labelme` (`https://github.com/wkentaro/labelme`) is a popular free software package for labeling images for segmentation tasks. It provides tools to mark various primitives: polygons, rectangles, circles, lines, and points. You can run the following command in a Terminal to install `labelme` using `pip`:

```
pip install labelme
```

This tool's usage is similar to `labelImg` introduced in *Chapter 3*. You can execute one of the following commands in a Terminal to start labeling:

```
labelme image_folder/
labelme image_folder/ --labels labels.txt
```

The first command specifies `image_folder`, where all labeling images are stored. If you wish, the second command allows specifying a list of labels (categories of objects to label) in a text file (`labels.txt`). The labeling process is then straightforward.

After labeling, there is one output text file per image. The output has the same filename but with a `.txt` extension. It is relatively easy to convert these annotation files into the COCO format supported by Detectron2. For instance, we can use the `labelme2coco` package (`https://github.com/fcakyon/labelme2coco`) for this task. The following command is the basic usage of this package:

```
labelme2coco dataset_path
```

The `dataset_path` argument is the path to the images and annotations labeled using `labelme`. Another useful usage pattern of this package is the following:

```
labelme2coco dataset_path --train_split_rate 0.8
```

The second argument (`--train_split_rate`) allows us to split the data into train and test sets with a specified ratio (`0.8` in this case).

The data sources and tools described in this section provide common data formats. However, in reality, some datasets are stored in different storage standards, and we might need to build the dataset in the format supported by Detectron2 from scratch. The next section introduces such a dataset and the steps to prepare this dataset.

Introduction to the brain tumor segmentation dataset

This chapter uses another brain tumor dataset but for segmentation tasks. This dataset is originally from *Jun Cheng's 2017 brain tumor dataset*: `https://doi.org/10.6084/m9.figshare.1512427.v5`. It has 3,064 images with 3 kinds of brain tumors: glioma (1,426 images), meningioma (708 images), and pituitary (930 images) tumors. The following section extracts the data, stores the extracted data in COCO format, and performs a train/test split.

Data extraction

This dataset is in MATLAB data format (`.mat` file). That is why it is a good example to demonstrate the steps to extract data from various forms and build a standard dataset accepted by Detectron2 from scratch. Specifically, each `.mat` file has the following fields:

- `label`: 1 for meningioma, 2 for glioma, and 3 for pituitary tumor
- `PID`: the patient identification
- `image`: the image data
- `tumorBorder`: the polygon coordinates of the tumor border

The following code snippet downloads the dataset and unzips it:

```
!wget --quiet https://figshare.com/ndownloader/
articles/1512427/versions/5 -O segbraintumors.zip
data_folder = "segbraintumors"
!unzip -q segbraintumors.zip -d {data_folder}

sets = ["1-766", "767-1532", "1533-2298", "2299-3064"]
sets = ["brainTumorDataPublic_"+_ for _ in sets]
```

```
!unzip -q {data_folder}/{sets[0]}.zip -d {data_folder}
!unzip -q {data_folder}/{sets[1]}.zip -d {data_folder}
!unzip -q {data_folder}/{sets[2]}.zip -d {data_folder}
!unzip -q {data_folder}/{sets[3]}.zip -d {data_folder}
```

Observably, there are four subfolders corresponding with four zip files in this dataset.

As discussed in *Chapter 3*, Detectron2 supports the COCO annotation format. The COCO annotation file has fields for info, licenses, categories, images, and annotations. The following code snippet creates a helper method that creates a JSON object with the data for the first three fields and empty lists for the last two:

```
def create_headers():
    data_json = {
    "info":{
        "description": "Desc"
    },
    "licenses":[{
        "id":1,
        "url":"https://creativecommons.org/licenses/by/4.0/",
        "name":"CC BY 4.0"
    }],
    "categories":[
        {"id":0,"name":"tumors","supercategory":"none"},
        {"id":1,"name":"meningioma","supercategory":"tumors"},
        {"id":2,"name":"glioma","supercategory":"tumors"},
        {"id":3,"name":"pituitary","supercategory":"tumors"}
        ],
    "images": [],
    "annotations":[]
    }
    return data_json
```

For the annotations, besides the segmentation masks (segmentation), there are bounding boxes (bbox) and areas (area). Therefore, the following code snippets create two helper methods to compute these fields from segmentation coordinates:

```
def segmentation2bbox(segmentation):
    segmentation = np.array(segmentation)
```

```
    segmentation = segmentation.reshape((-1, 2))
    x_min = segmentation[:, 0].min()
    y_min = segmentation[:, 1].min()
    x_max = segmentation[:, 0].max()
    y_max = segmentation[:, 1].max()
    width = x_max - x_min
    height= y_max - y_min
    bbox = [x_min, y_min, width, height]
    return bbox


def bbox2area(bbox):
    return bbox[2]*bbox[3]
```

Additionally, the following code snippet creates a helper method to generate JSON objects for images:

```
def create_image_obj(id, file_name, height, width):
    return {
        "id": id,
        "license": 1,
        "file_name": file_name,
        "height": height,
        "width": width
    }
```

Similarly, the next one creates a helper method to generate JSON objects for annotations:

```
def create_annotation_obj(id,
                          image_id,
                          category_id,
                          segmentation,
                          bbox):
    iscrowd = 0
    area = bbox2area(bbox)
    return {
        "id": id,
        "image_id": image_id,
        "category_id": int(category_id),
        "segmentation": [segmentation.tolist()],
```

```
        "bbox": bbox,
        "iscrowd": iscrowd,
        "area": area
    }
```

The following code snippet creates a JSON object (data_json) to store annotations, then loops through the .mat files and performs the following tasks on each file:

1. Extracts the tumor image data and writes the image into a file

2. Creates a JSON object for the image and appends it to data_json

3. Extracts planar coordinates (boundaries) and labels and creates JSON objects for the annotations, and appends them to data_json:

```
data_json = create_headers()
for i in tqdm(range(1, 3064+1)):
    with h5py.File(f'{data_folder}/{i}.mat', 'r') as f:
        obj = f['cjdata']
        # Step 1: extract image and write it to file
        # Step 2: create JSON object for image and append it
        # Step 3: extract boundaries + labels then append them
```

Step 1: This step extracts the image data and writes it into a file. Notably, its data type is float64 to keep the image quality:

```
image = obj['image'][:, :].astype('float64')
image = (image/image.max())*255.0
file_name = f"{i}.jpg"
cv2.imwrite(os.path.join(output_folder, file_name), image)
```

Step 2: This step creates the image JSON object and appends it into data_json:

```
height, width = image.shape[:2]
data_json["images"].append(
    create_image_obj(
        id        = i,
        file_name = file_name,
        height    = height,
        width     = width
))
```

Step 3: This step extracts labels and their corresponding planar coordinates for the brain tumors in the current `.mat` file. It then creates corresponding annotation objects and appends them to `data_json`:

```
label = obj['label'][:, :]
tumorBorder = obj['tumorBorder'][:, :]
for j, lbl in enumerate(label):
    segmentation = tumorBorder[j].reshape((-1, 2))[:, [1, 0]].
reshape((-1))
    bbox = segmentation2bbox(segmentation)
    data_json["annotations"].append(
        create_annotation_obj(
            id              = i,
            image_id        = i,
            category_id     = lbl[0],
            bbox            = bbox,
            segmentation    = segmentation
))
```

The following code snippet then writes the extracted annotations (`data_json`) into a JSON COCO annotation file:

```
af = "_annotations.coco.json"
with open(os.path.join(output_folder, af), "w") as f:
    json.dump(data_json, f)
```

The dataset is now in COCO format. However, it contains all the images and labels in one set. Therefore, the following section splits this dataset into `train` and `test` sets.

Train/test split

The following code snippet creates a helper method to write a JSON file given fields for the COCO annotation format (`info`, `licenses`, `categories`, `images`, and `annotations`):

```
def create_json(info,
                licenses,
                categories,
                images,
                annotations,
                file_name):
    obj = {
```

```
        "info"        : info,
        "licenses"    : licenses,
        "categories"  : categories,
        "images"      : images,
        "annotations" : annotations
    }
    with open(file_name, "w") as f:
      json.dump(obj, f)
    print(f"Saved {file_name}")
```

Next, we extract the COCO annotation file for the whole dataset and get the data for its fields:

```
name_ds = output_folder
af = "_annotations.coco.json"
with open(os.path.join(name_ds, af), "r") as f:
  annotations_json = json.load(f)

info        = annotations_json["info"]
licenses    = annotations_json["licenses"]
categories  = annotations_json["categories"]
images      = annotations_json["images"]
annotations = annotations_json["annotations"]
```

The first three fields (info, licenses, and categories) are the same for train and test sets. Therefore, we need to only do the splitting for the last two (images and annotations). In this case, one image has one annotation and is stored in the same order. Therefore, it is relatively easy to perform the split:

```
from sklearn.model_selection import train_test_split
stratify = [i['category_id'] for i in annotations]
test_size = 0.1
images_train, images_test, annotations_train, annotations_test
= train_test_split(
    images,
    annotations,
    test_size     = test_size,
    stratify      = stratify,
    random_state  = 42
)
```

We use the tumor type (`category_id`) as a stratified field to split tumor types between the `train` and `test` sets. The following code snippet helps to create `train` and `test` directories, move the images, and create the corresponding annotation files.

First, we create the corresponding folders for the `train` and `test` datasets:

```
train_path  = os.path.join(name_ds, "train")
test_path   = os.path.join(name_ds, "test")
train_af    = os.path.join(train_path, af)
test_af     = os.path.join(test_path, af)
os.makedirs(train_path, exist_ok=True)
os.makedirs(test_path, exist_ok=True)
```

Next, we move images to their corresponding `train` and `test` folders:

```
for img in tqdm(images_train):
  frm = os.path.join(name_ds, img["file_name"])
  to  = train_path
  shutil.move(frm, to)
for img in tqdm(images_test):
  frm = os.path.join(name_ds, img["file_name"])
  to  = test_path
  shutil.move(frm, to)
```

Finally, we write annotations to the corresponding `train` and `test` folders:

```
create_json(info,
            licenses,
            categories,
            images_train,
            annotations_train,
            file_name = train_af)
create_json(info,
            licenses,
            categories,
            images_test,
            annotations_test,
            file_name = test_af)
```

Optionally, you may want to run the following code snippet to remove the current annotation file of the whole dataset:

```
!rm {name_ds}/{af}
```

If you are running this notebook on Google Colab, it is time to zip this dataset before downloading it using the following snippet:

```
!zip -q -r {name_ds}.zip {name_ds}
```

Lastly, the following code snippet downloads this dataset to the local computer for future use:

```
from google.colab import files
files.download(f"{name_ds}.zip")
```

Congratulations! By this time, you should have mastered the components of the COCO annotation format and how to perform the necessary steps to extract data from various sources and form datasets in this annotation format from scratch. We are ready to train a custom segmentation application in Detectron2. However, we should first understand the architecture of such an application before training one.

The architecture of the segmentation models

Chapter 4 dived deep into the architecture of Faster R-CNN implemented in Detectron2 for object detection tasks in computer vision. Detectron2 implements Mask R-CNN for object segmentation tasks. It has the same components as Faster R-CNN architecture and one other head for the segmentation task. *Figure 10.1* shows the Faster R-CNN architecture as listed in *Chapter 4*.

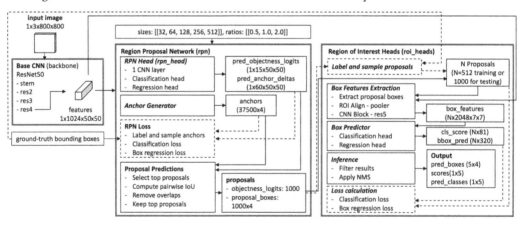

Figure 10.1: The architecture of the Detectron2 object detection application

The code to explore the architecture of Detectron2 implementation for object segmentation tasks remains the same as that for the object detection tasks. Therefore, this section only covers some discrepancies between the two. The first difference is that we must select a pre-trained model with configuration for the object segmentation task instead of the detection task. Therefore, we change the corresponding paths for the configuration and model weights as follows:

```
config_file = "<url_to/mask_rcnn_R_50_C4_1x.yaml>"
checkpoint_url = "<url_to/mask_rcnn_R_50_C4_1x.yaml>"
```

Once again, as discussed in *Chapter 2*, other models are available, and there are trade-offs between selecting one versus the others. This selection selects this model (Res50C4) due to its simplicity. The backbone and region proposal networks remain the same and work similarly for object detection tasks. The Region of Interest Heads (roi_heads) has one additional head called the mask head (mask_head). Therefore, the output of the following code snippet reflects the differences:

```
roi_heads = predictor.model.roi_heads
print(roi_heads)
```

This snippet, besides pooler, res5, and box_predictor as in *Chapter 4*, also displays the mask head as the following:

```
(mask_head): MaskRCNNConvUpsampleHead(
(deconv): ConvTranspose2d(2048, 256, kernel_size=(2, 2),
stride=(2, 2))
(deconv_relu): ReLU()
(predictor): Conv2d(256, 80, kernel_size=(1, 1), stride=(1, 1))
)
```

This mask head is an instance of MaskRCNNUpsampleHead (for this specific case). It is relatively simple, consisting of a ConvTranspose2d layer with ReLU activation and a Conv2d layer. This mask head is configurable via the name parameter in the configuration file (cfg.MODEL. ROI_MASK_HEAD.NAME).

Figure 10.2 shows the expanded Region Of Interest Heads with `mask_head` for object segmentation tasks.

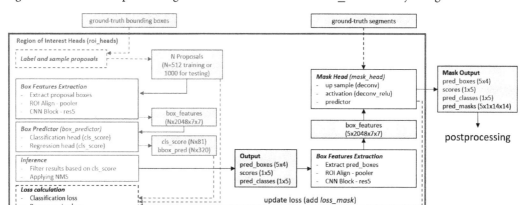

Figure 10.2: Region of Interest Heads expanded with the mask head for segmentation tasks

Specifically, all the inputs/outputs and intermediate processing are the same as those from the architecture for the object detection tasks (they are faded out). The extension part (not faded) utilizes the predicted bounding boxes from the detection task as its proposals. It then utilizes another **Box Features Extraction** component that works the same as in the bounding box prediction part. The only difference is that it uses the predicted boxes of the previous step as proposals instead of the original proposals from the region proposal network. Other than that, this component works the same way. The following code snippet shows how this component works:

```
# extract features from the predicted boxes
pred_boxes = [
    instance.pred_boxes
    for instance
    in pred_instances
    ]
x = roi_heads.pooler([res4], pred_boxes)
x = roi_heads.res5(x)
print(x.shape)
```

Chapter 4 provided details about the `pooler` and `res5` components. This code snippet should display the extracted features as a tensor of the size `torch.Size([5, 2048, 7, 7])`. In this case, the first value (5) means the number of boxes predicted in the previous step (detection task). Next, the extracted features, together with the predicted instances (outputs of the previous steps), are passed through the mask head:

```
output = mask_head(x, pred_instances)
```

Besides the fields from the instances predicted by the object detection task (`pred_boxes`, `pred_classes`, and `scores`), this output has a field for the predicted masks (`pred_masks`), one per predicted box. Additionally, if it is at the training stage, this mask head takes the ground-truth segmentations and updates the overall loss by adding the mask loss (`mask_loss`), as shown in *Figure 10.2*. The following code snippet displays the sizes of the predicted masks:

```
print(output[0].pred_masks.shape)
```

There is a corresponding predicted mask per predicted box. Therefore, this code snippet should display the tensor size of `torch.Size([5, 1, 14, 14])`. Again the first number (5) is the number of predicted bounding boxes. The mask resolution is configurable via the following parameter:

```
cfg.MODEL.ROI_MASK_HEAD.POOLER_RESOLUTION
```

This value is currently `14` (as the output mask resolution). The following code snippet visualizes these predicted masks:

```
import matplotlib.pyplot as plt
fig, axes = plt.subplots(1, 5, figsize=(15, 3))
for i in range(5):
    mask = output[0].pred_masks[i][0]
    mask = mask.detach().to("cpu")
    axes[i].imshow(mask)
plt.tight_layout()
plt.show()
```

This code snippet extracts the predicted masks, detaches them, and visualizes them. *Figure 10.3* shows the output of this code snippet.

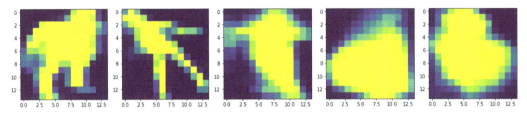

Figure 10.3: The masks (14x14 resolution) of the five predicted instances

These masks have distorted shapes of the detected instances because they have a squared resolution (14x14 in this case), while the predicted instances may have different ratios as in pred_boxes. Therefore, there is an extra step to scale these masks to the corresponding predicted bounding box sizes. The postprocessing step does this scaling:

```
from detectron2.modeling.postprocessing import detector_
postprocess
post_output = detector_postprocess(
    results       = output[0],
    output_height = 800,
    output_width  = 800,
    )
```

This code snippet takes the output instance and the original input image size (800 by 800 pixels) and scales all the predicted results into sizes corresponding to the image size. Additionally, the predicted masks are scaled to their corresponding bounding box sizes. Finally, the following code snippet visualizes the predicted results:

```
metadata = MetadataCatalog.get(cfg.DATASETS.TRAIN[0])
v = Visualizer(img[:, :, ::-1], metadata, scale=0.5)
instances = post_output.to("cpu")
annotated_img = v.draw_instance_predictions(instances)
imshow(annotated_img.get_image())
```

This visualization code snippet is the same as that for the object detection task, and it should display the output as shown in the center of *Figure 10.4*.

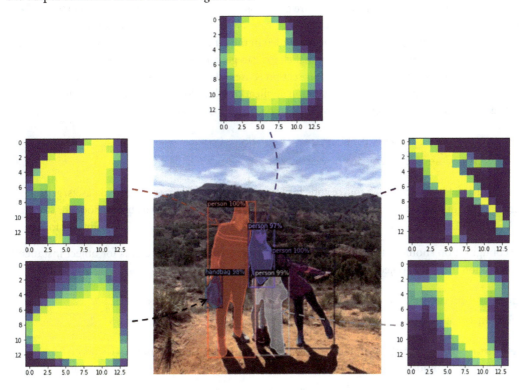

Figure 10.4: Sample output of an object segmentation application

The detected masks in the final output are scaled to the corresponding detected bounding boxes in the final output.

Congratulations! By this time, you should have a profound knowledge of the architecture of the Detectron2 instance segmentation application. Mainly, it has the same architecture as the object detection application with some extensions for predicting the masks using the predicted detection results as input. With the data ready and a deep understanding of the application architecture, you are ready to train an object segmentation model using Detectron2 in the following section.

Training custom segmentation models

The previous section described the architecture of an object segmentation application that extends the object detection application. Therefore, the training steps and configurations of these two applications are similar. This section utilizes the source code provided in the *Putting it all together* section in *Chapter 7*. Ideally, we should perform all the steps in that chapter to get a better set of hyperparameters for this specific dataset. However, for simplicity, this section reuses all the source code as it is and focuses on the differences for training object segmentation applications. The following code snippet downloads the dataset prepared in the previous steps with `train` and `test` sets:

```
!wget -q <url_to/segbraintumors_coco.zip>
!unzip -q segbraintumors_coco.zip
```

Once the dataset is extracted, the source code for Detectron2 installation, registering datasets, getting a training configuration, building a custom trainer with COCO Evaluator and a custom hook for storing the best model, training, and visualizations of the results is the same, except a few points, as follows.

The configuration should set the pre-trained model configuration and weights for object instance segmentation instead of object detection:

```
config_file_url = "<url_to/mask_rcnn_R_50_FPN_3x.yaml>"
checkpoint_url = "<url_to/mask_rcnn_R_50_FPN_3x.yaml>"
```

The `BestModelHook` class should use the segmentation evaluation metric (`segm/AP50`) for selecting the best model instead of the object detection evaluation metric (`bbox/AP50`):

```
class BestModelHook(HookBase):
  def __init__(self, cfg,
                metric  = "segm/AP50",
                min_max ="max"):
    # other lines remain the same
```

The preceding code snippet changes the default evaluation metric, and the following code snippet changes it when creating a new hook instance:

```
bm_hook = BestModelHook(cfg,
                        metric  = "segm/AP50",
                        min_max = "max")
```

The training then should proceed in the same way. However, besides the losses for object detection tasks – classification loss (loss_cls), bounding box regression loss (loss_box_reg), region proposal classification loss (loss_rpn_cls), and region proposal location loss (loss_rpn_loc) – there is one more loss metric for the masks (loss_mask) for instance segmentation tasks:

```
total_loss: 1.325  loss_cls: 0.1895  loss_box_reg:
0.1189  loss_mask: 0.5053  loss_rpn_cls: 0.3287  loss_rpn_loc:
0.1724
```

The preceding listing is one example of the losses during the training object instance segmentation model. Additionally, at every evaluation iteration, besides the evaluation metrics for the bounding box (bbox), there is a similar set of metrics for segmentation (segm), which is used for selecting the best model:

```
Evaluation results for segm:
|   AP   |  AP50  |  AP75  |  APs   |  APm   |  APl   |
|:------:|:------:|:------:|:------:|:------:|:------:|
| 25.786 | 41.874 | 27.180 | 4.064  | 26.409 | 34.991 |
```

The following code snippet sets the confidence threshold as 0.7 and loads the best model:

```
cfg.MODEL.ROI_HEADS.SCORE_THRESH_TEST  = 0.7
cfg.MODEL.WEIGHTS = os.path.join(cfg.OUTPUT_DIR,
                              "model_best.pth")
predictor = DefaultPredictor(cfg)
```

The following code snippets create some helper methods to randomly predict and visualize three data samples and display the ground-truth masks versus the predicted masks for qualitative evaluation purposes. Specifically, the following code snippet creates a method (plot_samples) that takes samples as a data dictionary and predictor and then visualizes the ground truths or predicted results depending on the is_gt input:

```
def plot_samples(samples,
                  met        = {},
                  is_gt      = True,
                  predictor  = None):
  n = len(samples)
  nrows, ncols = int(-(-n/3)), 3
  fig, axs = plt.subplots(nrows, ncols, figsize = (21, 7))
  for i,s in enumerate(samples):
    row, col = i//ncols, i%ncols
```

```
      ax = axs[row][col] if len(axs.shape)==2 else axs[i]
      img = cv2.imread(s["file_name"])
      v = Visualizer(img[:,:, ::-1], metadata=met, scale=0.5)
      if is_gt:    # ground-truth
        v = v.draw_dataset_dict(s)
      else:     # predictions
        instances = predictor(img)["instances"].to("cpu")
        v = v.draw_instance_predictions(instances)
      ax.imshow(v.get_image())
      ax.axis("off")
    plt.tight_layout()
    plt.show()
```

The next method (plot_random_samples) samples some inputs from a registered dataset (name_ds) and plots the ground truths and the predicted results for qualitative evaluation purposes:

```
  def plot_random_samples(name_ds, n=3, predictor=None):
    ds = DatasetCatalog.get(name_ds)
    met = MetadataCatalog.get(name_ds)
    samples = random.sample(ds, n)
    # plot samples with ground-truths
    plot_samples(samples, met)
    # plot predictions
    plot_samples(samples,
                    met          = met,
                    predictor    = predictor,
                    is_gt        = False)
```

All in all, these two code snippets create two helper methods. The first method helps to visualize either a ground-truth data dictionary or to predict output given the data dictionary and the model and visualize the results. The second method randomly samples for a number (n) of samples and passes them to the first method to visualize. The following code snippet utilizes the created method to visualize some results for qualitative evaluation purposes:

```
  plot_random_samples(name_ds_test, predictor = predictor)
```

Figure 10.5 shows the sample output of the ground truth masks (top line) and the predicted results (bottom line).

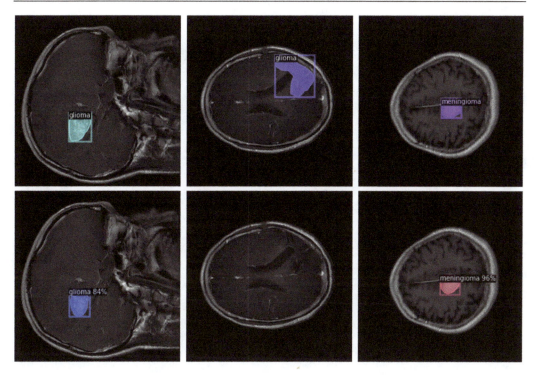

Figure 10.5: Sample prediction results from object instance segmentation model

The predicted results are good with a simple model and configuration, even though it still missed one prediction. Further training, better optimizations, and more complicated architecture should help improve the results.

Congratulations! You have now mastered the steps to train an object instance segmentation application on a custom dataset using Detectron2.

Summary

This chapter described the common places to acquire data for object instance segmentation tasks, and the steps to extract data from a non-standard annotation format and create a custom dataset in the format supported by Detectron2. Furthermore, it utilized the code and visualizations approach to illustrate the architecture of the object instance segmentation application implemented in Detectron2 as an extension of the architecture of an object detection application. Finally, it provided the steps to train a custom model for object instance segmentation tasks and visualize the prediction results for qualitative evaluations.

The next chapter describes techniques for fine-tuning the object instance segmentation application and improving segmentation quality using Detectron2.

11

Fine-Tuning Instance Segmentation Models

The object instance segmentation models utilize results from the object detection models. Therefore, all the techniques introduced in the previous chapters for fine-tuning object detection models work the same for object instance segmentation models. However, object instance segmentation has an important feature to fine-tune: the quality of the boundaries of the detected objects. Therefore, this chapter introduces **PointRend**, a project inside Detectron2 that helps improve the object boundaries' sharpness.

By the end of this chapter, you will be able to understand how PointRend works. You will also have hands-on experience developing object instance segmentation applications with better segmentation quality using existing PointRend models. Additionally, you can train an object instance segmentation application using PointRend on a custom dataset. Specifically, this chapter covers the following topics:

- Introduction to PointRend
- Using existing PointRend models
- Training custom PointRend models

Technical requirements

You must have completed *Chapter 1* to have an appropriate development environment for Detectron2. All the code, datasets, and results are available in this book's GitHub repository at https://github. com/PacktPublishing/Hands-On-Computer-Vision-with-Detectron2.

Introduction to PointRend

PointRend is a project as a part of Detectron2. It helps provide better segmentation quality for detected object boundaries. It can be used with instance segmentation and semantic segmentation. It is an extension of the Mask R-CNN head, which we discussed in the previous chapter. It performs point sampling on the detected mask and performs predictions on the sampled points instead of all the points in the mask. This technique allows us to compute the mask with a higher resolution, thus providing a higher mask resolution. *Figure 11.1* illustrates an example of two images when not using (left) and when using (right) PointRend:

Without PointRend   With PointRend

Figure 11.1: Segmentation quality with and without PointRend

PointRend helps render segmentations that are of a higher resolution with object boundaries that are crisp and less smooth. Therefore, it is useful if the objects to detect have sharp edges.

At inference time, starting from the coarse prediction, it performs upsampling via several steps to increase the mask resolution. At each step, there are three main tasks:

1. Sample a number (hyperparameter) of points for performing predictions. This strategy helps avoid produced computing predictions on many grid points at higher resolutions.

2. Extract features for each sampled point from the feature maps extracted by the backbone, combined with the features from the coarse prediction.

3. Classify each sampled point's labels using their extracted features using a multilayer perceptron neural network (Point Head).

Figure 11.2 illustrates the PointRend strategy in three different resolutions for one object instance at inference time:

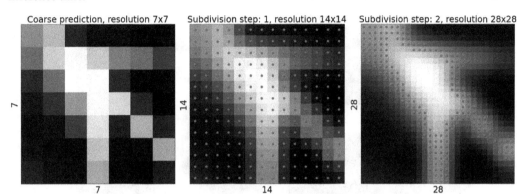

Figure 11.2: PointRend prediction at inference time

The leftmost image is the *7×7* coarse prediction discussed in the previous chapter. PointRend has a hyperparameter for the number of points to sample ($N = 14×14$, in this case); thus, the coarse prediction is upsampled (for example, using bilinear interpolation), as in the middle of the preceding figure. This solution has the same number of pixels as the number of points to sample. Thus, every single point is passed to the Point Head to predict its label (middle of the figure). In the next subdivision step, the image is upsampled to a *28×28* resolution. This resolution has more pixels than the N points. Therefore, PointRend applies a sample strategy here. The N sampled points are those more ambiguous (in terms of whether or not they should be in the object, and they represent the boundary). In other words, these *14×14* points sampled to be passed to the Point Head to perform prediction have prediction probabilities close to *0.5* (ambiguous). The reason for this is obvious: those with high confidence in being in the background or being in the object do not require further predictions to be made. This approach, at inference time, is iterative.

This iterative approach is not friendly for training neural networks. Therefore, PointRend utilizes a different point sampling strategy at training time to train its Point Head. This sampling strategy has three parameters: the number of points to sample (N), the upsampling constant (k), and the importance sample ratio (β), which is the percentage of points that are closer to the edges (ambiguous or having a prediction confidence close to *0.5*). Specifically, these points are selected via the following steps:

1. Sample for kN points uniformly in the coarse predictions (sampling for kN instead of N makes this step an oversampling step and k as the oversampling constant, where $k > 1$).

2. Select βN points closest to the boundary (ranked by how close their prediction confidences are to *0.5*). These points are closer to the edge, and that makes them important. Therefore, β is called the importance sample ratio.

3. Select $(1-\beta)N$ points from the kN points sampled in *step 1* to have a total of N points for training the Point Head.

Figure 11.3 illustrates an example of the selected points (right) for the coarse mask prediction of the *7×7* resolution (left) with *N = 14×14, k = 3*, and *β = 0.75*:

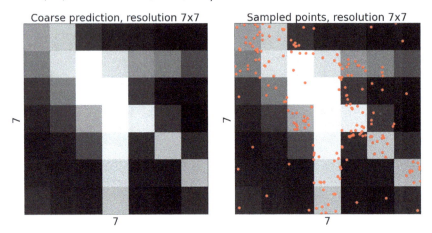

Figure 11.3: PointRend point sampling strategy at training time

More points are selected (approximately 75% are at the boundary) and the rest are chosen uniformly randomly over the coarse mask to extract features and train the Point Head.

Like other Detectron2 models, the PointRend project provides a set of built-in models for its Model Zoo, which can be accessed via the Detectron2 GitHub repository under the `projects/PointRend` folder. The same analysis that we used in *Chapter 2* can be used to select an appropriate PointRend model with the trade-off of accuracy versus inference time. The next section demonstrates the steps for creating an object instance segmentation application using existing models from the PointRend Model Zoo and illustrates how these steps work.

Using existing PointRend models

The steps for performing object instance segmentation using existing PointRend models are similar to that of performing object instance segmentation using existing Detectron2 models in the Detectron2 Model Zoo, as described in the previous chapter. Therefore, this section covers more of the differences. For PointRend, we need to clone the Detectron2 repository to use its configuration files from the PointRend project:

```
!git clone --branch https://github.com/facebookresearch/
detectron2.git detectron2_repo
```

The repository is stored in the `detectron2_repo` folder in the current working directory. With this repository cloned, the code to generate the configuration is a little different:

```
# some other common import statements are removed here
from detectron2.projects import point_rend
config_file = "detectron2_repo/projects/PointRend/configs/
InstanceSegmentation/pointrend_rcnn_X_101_32x8d_FPN_3x_coco.
yaml"
checkpoint_url = "detectron2://PointRend/InstanceSegmentation/
pointrend_rcnn_X_101_32x8d_FPN_3x_coco/28119989/model_final_
ba17b9.pkl"
cfg = get_cfg()
point_rend.add_pointrend_config(cfg)
cfg.merge_from_file(config_file)
cfg.MODEL.WEIGHTS = checkpoint_url
score_thresh_test = 0.95
cfg.MODEL.ROI_HEADS.SCORE_THRESH_TEST = score_thresh_test
predictor = DefaultPredictor(cfg)
```

Specifically, this code snippet utilizes the existing configuration file and model weight selected from the PointRend Model Zoo. It also adds PointRend-specific configuration using the files cloned from the Detectron2 repository. The rest of the code for generating the configuration file, downloading the existing weights, and creating a predictor using `DefaultPredictor` is similar.

The following code snippet creates a helper method to visualize a given image in its original size:

```
def imshow(image):
    dpi = plt.rcParams["figure.dpi"]
    im_data = image[:,:, ::-1]
    height, width, depth = im_data.shape
    figsize = width / float(dpi), height / float(dpi)
    fig = plt.figure(figsize=figsize)
    plt.imshow(im_data)
    plt.imshow(im_data)
    plt.axis("off")
    plt.show()
```

The following code utilizes the `imshow` method and creates a method to visualize an image, given a configuration object and the prediction output:

```
def visualize(img, cfg, output):
  metadata = MetadataCatalog.get(cfg.DATASETS.TRAIN[0])
  v = Visualizer(img[:, :, ::-1], metadata, scale=1.0)
  instances = output["instances"].to("cpu")
  annotated_img = v.draw_instance_predictions(instances)
  imshow(annotated_img.get_image()[:, :, ::-1])
```

The following code snippet performs the prediction and visualizes the results on a test image (`img`):

```
output = predictor(img)
visualize(img, cfg, output)
```

Figure 11.4 shows the output of this code snippet:

Figure 11.4: Example of object instance segmentation results using PointRend

Observably, the object instance segmentation boundaries predicted by PointRend models are crisp and have better quality compared to the object instance segmentation boundaries produced using standard object instance segmentation models in Detectron2. You can refer to *Figure 11.1* again to compare the quality of the segmentation boundaries between these two versions.

Using existing models on the PointRend Model Zoo can meet the most common needs with models trained on popular datasets for computer vision tasks. However, if needed, PointRend allows us to train its models on custom datasets too. The next section describes the steps to train PointRend models on a custom dataset.

Training custom PointRend models

This section describes the steps for training a custom PointRend model for object instance segmentation tasks on the brain tumor dataset (described in the previous chapter). Training custom PointRend models includes steps similar to training instance segmentation models in Detectron2. Therefore, this section focuses more on the discrepancies between the two; the complete source code is available in this book's GitHub repository.

The source code for downloading the brain tumor segmentation dataset, installing Detectron2, and registering train and test datasets remains the same. Similar to the previous section, before we can get a configuration file, we need to clone the Detectron2 repository to use the configuration files for the PointRend project:

```
!git clone https://github.com/facebookresearch/detectron2.git
detectron2_repo
```

The code for the initial configuration remains the same:

```
output_dir = "output/pointrend"
os.makedirs(output_dir, exist_ok=True)
output_cfg_path = os.path.join(output_dir, "cfg.pickle")
nc = 3
device = "cuda" if torch.cuda.is_available() else "cpu"
```

Additionally, the following code snippet is similar to the one shown in the previous section for selecting an existing PointRend model on the PointRend Model Zoo:

```
from detectron2.projects import point_rend
config_file = "detectron2_repo/projects/PointRend/configs/
InstanceSegmentation/pointrend_rcnn_R_50_FPN_3x_coco.yaml"
checkpoint_file = "detectron2://PointRend/InstanceSegmentation/
pointrend_rcnn_R_50_FPN_3x_coco/164955410/model_final_edd263.
pkl"
cfg = get_cfg()
point_rend.add_pointrend_config(cfg)
cfg.merge_from_file(config_file)
cfg.MODEL.WEIGHTS = checkpoint_file
```

For the next few sections of code for the configuration file, ideally, it would be better to run steps to generate different configurations for learning rates, pixel means and standard deviations, and anchor sizes and ratios. However, for simplicity, we have reused the results from previous executions. Additionally, the PointRend model requires a setting for the number of classes for its Point Head; the following code snippet sets that configuration:

```
cfg.MODEL.POINT_HEAD.NUM_CLASSES = nc
```

Once we have the necessary configuration, we are ready to train the model. The code snippets for creating the trainer class (`BrainTumorTrainer`), the best model hook (`BestModelHook`), creating a trainer, and registering the hook remain the same as the source code explained in the previous chapter.

The current PointRend project, at the time of writing, throws an error during training if a training batch does not produce any instances (an empty instance list). This issue does not occur for most of the datasets but would occur for some. For example, this issue happens for the training process of the current model trained on the brain tumor segmentation dataset due to its small batch size. Therefore, the following code snippet modifies the PointRend internal sampling strategy code to handle this exception. The following code snippet imports the packages required for this task:

```
import torch
from detectron2.layers import cat
from detectron2.structures import Boxes
from detectron2.projects.point_rend.point_features import(
    get_point_coords_wrt_image,
    sample_point_labels,
    get_uncertain_point_coords_with_randomness
)
from detectron2.projects.point_rend.mask_head import (
    calculate_uncertainty
)
```

The following code snippet redefines the _sample_train_points method to handle the exception:

```
def _sample_train_points(self, coarse_mask, instances):
    assert self.training
    gt_classes = cat([x.gt_classes for x in instances])
    with torch.no_grad():
        # Step 1: sample point coordinates (point_coords)
        # Step 2: get point_coords with respect to the image
```

```
    # Step 3: generate labels for the sampled points
    return point_coords, point_labels
```

This method has three main parts:

1. It samples the point coordinates with some randomness.
2. It converts the sampled point coordinates into the coordinates concerning the original image.
3. It gets the labels for the sampled point coordinates.

The first part of the method (*steps 1* and *2*) remains the same as the original method in the `PointRendMaskHead` class from the PointRend project (inside the `mask_head.py` file). The changes are placed at the end (*step 3*). Specifically, the changes include creating a default tensor for the point labels as zeros, trying to perform the sampling process, and printing an error message if an exception occurs. Let's look at the code for these three steps.

Step 1: This step samples the point coordinates with some randomness:

```
point_coords=get_uncertain_point_coords_with_randomness(
    coarse_mask,
    lambda logits: calculate_uncertainty(
        logits,
        gt_classes),
    self.mask_point_train_num_points,
    self.mask_point_oversample_ratio,
    self.mask_point_importance_sample_ratio,
)
```

Step 2: This step converts the sampled point coordinates into the coordinates of the original image:

```
proposal_boxes = [x.proposal_boxes for x in instances]
cat_boxes = Boxes.cat(proposal_boxes)
point_coords_wrt_image = get_point_coords_wrt_image(
    cat_boxes.tensor,
    point_coords)
```

Step 3: This step generates the labels for the sampled points. This step is what requires us to reimplement the original method. This step is different from the original method. First, it generates labels with all default values set to zeros. It then tries to create the labels as in the original implementation. If errors occur, it displays an exception message, and the default labels are used:

```
sR, sP, s2 = point_coords.shape
assert s2 == 2, point_coords.shape
point_labels = torch.zeros( size    = (sR, sP),
                            dtype   = point_coords.dtype,
                            layout  = point_coords.layout,
                            device  = point_coords.device)
try:
  point_labels = sample_point_labels(
      instances,
      point_coords_wrt_image)
except:
  print("*************empty instances*************")
```

We do not wish to modify the original PointRend source code; instead, we wish to modify this method on the current trainer object. Therefore, the following code snippet utilizes the `types` package to make this change:

```
import types
mask_head = trainer.model.roi_heads.mask_head
mask_head._sample_train_points = types.MethodType(
    _sample_train_points,
    mask_head
)
```

Now, we are ready to train the model:

```
trainer.train()
```

Besides other loss types described in the previous chapter, there is an additional loss called `loss_mask_point` during training. Additionally, you may sometimes see `*************empty instances*********` messages for the batches that produce empty lists of instances during training.

After the training completes, the source code for loading the best model and visualizing some random inputs from the test sets and their corresponding prediction outputs remains the same as what was listed in the previous chapter. *Figure 11.5* shows an example of a ground-truth segment (left) and the predicted one (right) while using our model trained with PointRend:

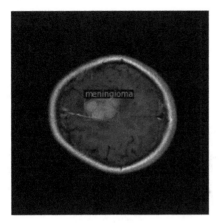 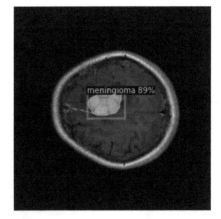

Ground-truth segment   Predicted segment using PointRend

Figure 11.5: The ground-truth segment versus the predicted one using PointRend

Observably, the predicted segment that uses PointRend (right) has a more precise boundary compared to the ground-truth one.

Congratulations! By now, you should have a profound understanding of the PointRend project and how it helps improve object instance segmentation quality. Additionally, you can develop object instance segmentation applications using existing PointRend models or by training PointRend models on custom datasets.

Summary

This chapter introduced the techniques to fine-tune object instance segmentation applications trained using Detection2. In general, object instance segmentation applications also use object detection parts. Therefore, all the methods that are utilized to fine-tune object detection models can be used for object instance segmentation models. Additionally, this chapter discussed the PointRend project, which helps improve the object instance segmentation boundaries for the detected objects. Specifically, it described how PointRend works and the steps for developing object instance segmentation applications using the existing models available in the PointRend Model Zoo. Finally, this chapter also provided code snippets to train custom PointRend models on custom datasets.

Congratulations again! By now, you should have profound knowledge regarding Detectron2 and be able to develop computer vision applications by using existing models or training custom models on custom datasets. The next chapter discusses the steps for bringing the built models into deployment.

Part 4:
Deploying Detectron2 Models into Production

This last part walks you through the steps in an export process to convert Detectron2 models into deployable artifacts. Specifically, it describes the standard file formats of deep learning models such as TorchScript and corresponding runtimes for these formats, such as PyTorch and C++ runtimes. It then provides the steps to convert Detectron2 models to the standard file formats and deploy them to the corresponding runtimes. Additionally, this part introduces Open Neural Network Exchange (ONNX) framework. This framework helps share deep neural networks across multiple frameworks and platforms. It is extremely helpful when deploying Detectron2 models into browsers or mobile environments is needed. Finally, this part describes D2Go, a framework for training, quantizing, and deploying models with minimal memory storage and computation requirements. Models created using this framework are extremely helpful for deploying into mobile or edge devices.

The fourth part covers the following chapters:

- *Chapter 12, Deploying Detectron2 Models into Server Environments*
- *Chapter 13, Deploying Detectron2 Models into Browsers and Mobile Environments*

12
Deploying Detectron2 Models into Server Environments

This chapter walks you through the steps of the export process to convert Detectron2 models into deployable artifacts. Specifically, it describes the standard file formats of deep learning models such as TorchScript and the corresponding runtimes for these formats, such as PyTorch and C++. This chapter then provides the steps to convert Detectron2 models to the standard file formats and deploy them to the corresponding runtimes.

By the end of this chapter, you will understand the standard file formats and runtimes that Detectron2 supports. You can perform steps to export Detectron2 models into TorchScript format using tracing or scripting method. Additionally, you can create a C++ application to load and execute the exported models.

In this chapter, we will cover the following topics:

- Supported file formats and runtimes for PyTorch models
- Deploying custom Detectron2 models

Technical requirements

You should have completed *Chapter 1* to have an appropriate development environment for Detectron2. All the code, datasets, and results are available on the GitHub page of the book at `https://github.com/PacktPublishing/Hands-On-Computer-Vision-with-Detectron2`.

Supported file formats and runtimes

This section introduces the development environments, file formats, and runtimes for PyTorch applications in general and Detectron2 applications specifically. It then uses simple examples to explain these concepts.

Development environments, file formats, and runtimes

There are three main concepts to grasp while developing and deploying deep learning models: the development environment and programming languages, the model file formats, and the runtime environments. *Figure 12.1* illustrates these three concepts and examples of each.

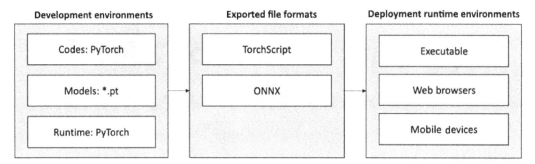

Figure 12.1: Development environments, file formats, and deployment environments

The development environment and programming languages or frameworks such as PyTorch or Python, in general, prioritize ease of use and customizations. However, the Python language and PyTorch framework may have runtime issues at deployment. For instance, they are relatively slow compared to other runtime environments, such as executable files generated by C++ applications. Additionally, they are not deployable directly on the web browser or mobile devices. Therefore, models written in PyTorch often need to go through an export process to become deployable artifacts.

Models written in PyTorch have their own file format and are meant to be executable with PyTorch runtime. However, to support cross-platform and compatibility with other runtime environments, PyTorch supports another intermediate file format called **TorchScript**. This TorchScript file can be executed in different runtime environments, such as being loaded by C++ applications to generate executable files. PyTorch also allows exporting its models into **Open Neural Network Exchange (ONNX)** file format to support other runtime environments such as web browsers and mobile devices.

TorchScript has its own interpreter, which functions like a limited version of Python. This format enables us to save the entire model to disk and use it in a different environment, such as a server built with another language other than Python. The TorchScript format allows for compiler optimizations, resulting in more efficient execution. Additionally, TorchScript can interface with multiple backend/device runtimes.

Detectron2 supports exporting its models into TorchScript using the `tracing` and `scripting` methods. It also allows users to export their models to an ONNX file format, which can be loaded into different runtime environments such as web browsers and mobile devices. Detectron2 models can be exported into Caffe2 file format using the `caffe2_tracing` method. However, Caffe2 is about to be deprecated and combined with PyTorch. Therefore, this book does not consider this option. Finally, TensorFlow is considered more popular in production environments compared to PyTorch at the time of writing. Therefore, Detectron2 supports exporting its models to TensorFlow's file format to execute them in the TensorFlow runtime environment. However, the conversion from Detectron2 models to TensorFlow is still limited and, thus, not covered in this book.

The following sub-sections use simple examples to illustrate the code for exporting PyTorch models into TorchScript models using tracing and scripting methods and deploying models into a C++ environment. The next section provides the code to export custom Detectron2 models into TorchScript.

Exporting PyTorch models using the tracing method

Let us create a simple PyTorch model to illustrate the export process. The model is simple enough to illustrate the point and allows us to focus on the code for exporting rather than building and training the models:

```
class SimplePyTorchModel(nn.Module):
  def __init__(self):
    super(SimplePyTorchModel, self).__init__()
    self.linear = nn.Linear(4, 1)
    self.linear.weight.data.fill_(0.01)
    self.linear.bias.data.fill_(0.01)


  def forward(self, x):
    return torch.relu(self.linear(x))


pt_model = SimplePyTorchModel()
```

This model (pt_model) has only one linear layer. Specifically, it only has four weights and one bias as parameters (all for the linear layer). Notably, we do not need to train this model but use 0.01 for all parameters for demonstration purposes.

Let us prepare some dummy data and perform a prediction. This predicted result can be compared with the future predictions of the exported models:

```
X = torch.tensor(
    data  = [[1, 2, 3, 4],
            [-1, -2, -3, -4]],
    dtype = torch.float32)
X = X.to(device)
with torch.no_grad():
  y = pt_model(X)
  print(y)
```

With the given parameters, this code snippet should display the following output:

```
tensor([[0.1100],
        [0.0000]], device='cuda:0')
```

The tracing technique passes a dummy input through the network and records the processing graph as the flow of the input from the start to the end of the model. The following code snippet prepares a dummy input with the same input size and traces the model (using the torch.jit.trace method):

```
dummy_X = torch.rand((1, 4))
dummy_X.to(device)
traced_model = torch.jit.trace(
    func            = pt_model.to("cpu"),
    example_inputs  = dummy_X)
traced_model.to(device)
```

This code snippet should display the traced model as follows:

```
SimplePyTorchModel(
  original_name=SimplePyTorchModel
  (linear): Linear(original_name=Linear)
)
```

We can also view the parameters of the traced models with the following code snippet:

```
for p in traced_model.parameters():
  print(p.data)
```

This code snippet should display the initial parameters assigned previously (0.01):

```
tensor([[0.0100, 0.0100, 0.0100, 0.0100]], device='cuda:0')
tensor([0.0100], device='cuda:0')
```

Executing the print(traced_model.code) statement allows us to view the code of this traced model as follows:

```
def forward(self,
    x: Tensor) -> Tensor:
  linear = self.linear
  return torch.relu((linear).forward(x, ))
```

Specifically, it has the code from the `forward` method of the original model in TorchScript format. Next, we are ready to perform the prediction using the traced model on the same input (X) created previously:

```
with torch.no_grad():
    y = traced_model(X)
    print(y)
```

The preceding code snippet should display the result exactly the same as the prediction of the original PyTorch model:

```
tensor([[0.1100],
        [0.0000]], device='cuda:0')
```

The following code snippet saves the traced model, loads it again, then uses it to perform prediction on the same input to make sure that we can export and load the model in a different environment:

```
traced_model.save("traced_model.pt")
loaded_model = torch.jit.load("traced_model.pt")
for p in loaded_model.parameters():
    print(p)
print(loaded_model(X))
```

This code snippet should save the traced model to the hard drive. After loading, the loaded model should have the same parameters as the PyTorch model. Finally, the loaded model should produce the same prediction on the X input as the original model.

If you run this notebook on Google Colab, you can save this model into your local system for future use with the following statements:

```
from google.colab import files
files.download('traced_model.pt')
```

Congratulations! Using the tracing method, you should have mastered the steps to export PyTorch models into TorchScript. The following section describes the limitation of the tracing method and introduces the scripting approach to perform the same task.

When the tracing method fails

The tracing method is simple and efficient. However, in several cases where your model uses control flows or loops, it may be necessary to write the model using TorchScript and annotate it accordingly. Similar to the previous section, let us use a simple model with default weights and bias as 0.01, and that has some control flows:

```
class SimplePyTorchModel(nn.Module):
  def __init__(self):
    super(SimplePyTorchModel, self).__init__()
    self.linear = nn.Linear(4, 1)
    self.linear.weight.data.fill_(0.01)
    self.linear.bias.data.fill_(0.01)

  def forward(self, x):
    y = self.linear(x)
    if y.sum() > 0:
      return y
    else:
      return -y

pt_model = SimplePyTorchModel()
```

Let us use a simple input and perform a prediction using this model:

```
X = torch.tensor(
    data  = [[1, 2, 3, 4]],
    dtype = torch.float32)
X = X.to(device)
with torch.no_grad():
  y = pt_model(X)
  print(y)
```

The prediction result is as follows:

```
tensor([[0.1100]], device='cuda:0')
```

Next, we can use the same code snippet as in the previous section to perform the export using the tracing method:

```
dummy_X = torch.tensor(
    data  = [[-1, -2, -3, -4]],
    dtype = torch.float32)
traced_model = torch.jit.trace(
    pt_model,
    example_inputs = dummy_X)
```

This code snippet should display a warning message with the following content:

```
TracerWarning: Converting a tensor to a Python boolean might
cause the trace to be incorrect. We can't record the data flow
of Python values, so this value will be treated as a constant
in the future. This means that the trace might not generalize
to other inputs!
```

This warning message indicates that the tracing approach only records the one/static flow of the input that passes through the model and does not consider other cases. You can use the same code snippet provided in the previous section to display the parameters' of the traced models:

```
traced_model.to(device)
for p in traced_model.parameters():
    print(p.data)
```

The parameters should remain the same (all 0.01 s). However, the model code is different. Specifically, print(traced_model.code) displays the following output:

```
def forward(self,
    x: Tensor) -> Tensor:
    linear = self.linear
    return torch.neg((linear).forward(x, ))
```

Note that this traced model always returns the negative (torch.neg) of the output from the linear model, as the path was taken by the dummy input and recorded in the traced model. In other words, the if code block will never be taken for future data.

Let us use the traced model and perform a prediction on the X input generated previously:

```
with torch.no_grad():
  y = traced_model(X)
  print(y)
```

This code snippet gives the following output:

```
tensor([[-0.1100]],device='cuda:0')
```

Observably, this is a different output compared to the prediction of the original PyTorch model because the execution path is for the dummy input (dummy_X) and not this current X input. Therefore, in this case, the scripting approach should be used.

Exporting PyTorch models using the scripting method

Continuing with the code in the previous section; the following code snippet exports the PyTorch model to TorchScript using the scripting method:

```
scripted_model = torch.jit.script(pt_model)
print(scripted_model.code)
```

This snippet should display the TorchScript code for the exported model as follows:

```
def forward(self,
    x: Tensor) -> Tensor:
  linear = self.linear
  y = (linear).forward(x, )
  if bool(torch.gt(torch.sum(y), 0)):
    _0 = y
  else:
    _0 = torch.neg(y)
  return _0
```

Observably, this is different from the tracing approach as the TorchScript code of the exported model now contains the control flow section. Additionally, the following code snippet should display the parameters of the exported model (as 0.01 for all parameters):

```
for p in scripted_model.parameters():
  print(p.data)
```

Furthermore, we can use the exported model to perform a prediction on the X input prepared previously to inspect the result:

```
scripted_model.to(device)
with torch.no_grad():
    y = scripted_model(X)
    print(y)
```

The prediction result should be exactly the same as what was produced by the original model (tensor([[0.1100]], device='cuda:0')). Additionally, we can save the model to the hard drive, load it back, display the parameter values for the loaded models, perform the prediction, and verify the result again:

```
# save
scripted_model.to("cpu").save("scripted_model.pt")
# load
loaded_model = torch.jit.load("scripted_model.pt")
# loaded parameters
for p in loaded_model.parameters():
    print(p)
# prediction
loaded_model.to(device)
loaded_model(X)
```

You may also download these models for future use if you are running this notebook on Google Colab:

```
from google.colab import files
files.download('scripted_model.pt')
files.download('pt_model.pt')
```

The tracing and scripting approaches have their strengths and weaknesses. The following section introduces the mixed approach to leverage the best of both.

Mixing tracing and scripting approaches

The tracing approach is simple and straightforward but does not properly record the models containing control flows and loops. The scripting approach can mitigate this limitation of the tracing approach. However, its supported syntax/code might be limited. Therefore, combining the two approaches may be beneficial in several cases.

For instance, tracing may be more appropriate in some instances than scripting, such as when a module has numerous design choices based on constant Python values and should not appear in TorchScript. Scripting is required when exporting models that contain control flows or loops. In these cases, scripting can be combined with tracing. The scripting (`torch.jit.script`) function produces the modules (codes) used by models that are then exported by the tracing approach. On the other hand, the tracing (`torch.jit.trace`) function can produce modules for models that are then exported by the scripting approach.

Congratulations! You have masted the two main techniques to export PyTorch models to TorchScript, which can be deployed in environments with server languages other than Python.

Deploying models using a C++ environment

PyTorch models are meant to be executed in a Python environment. Python can offer dynamicity, ease of use, and accommodate changes, and these features are useful at development time. However, Python might be unfavorable at deployment time due to slower execution performance. Deployment time often requires an execution environment with low latencies, such as C++. This section provides the code to load the exported TorchScript model to a C++ environment and performs a simple prediction to validate the result.

LibTorch is the C++ library for PyTorch. LibTorch is designed to be a fast and memory-efficient environment to execute deep learning models with C++. This makes it a good choice for real-time and performance-critical applications, where the ability to use hardware acceleration and custom hardware is important. Additionally, LibTorch can be used as a standalone library, independent of the PyTorch Python library. This feature enables the flexibility to use the best tool for deployment. The following code snippet installs LibTorch on Google Colab:

```
!wget --quiet <url_to/libtorch-shared-with-deps-latest.zip>
!unzip -qq libtorch-shared-with-deps-latest.zip
```

Next, we make a folder (cppapp) for the code of our C++ application:

```
import os
os.makedirs('cppapp', exist_ok=True)
```

The following code snippet uses the `%%writefile` directive to write a file for this C++ application (CMakeLists.txt):

```
%%writefile cppapp/CMakeLists.txt
cmake_minimum_required(VERSION 3.0 FATAL_ERROR)
project(cppapp)
find_package(Torch REQUIRED)
set(CMAKE_CXX_FLAGS "${CMAKE_CXX_FLAGS} ${TORCH_CXX_FLAGS}")
```

```
add_executable(cppapp cppapp.cpp)
target_link_libraries(cppapp "${TORCH_LIBRARIES}")
set_property(TARGET cppapp PROPERTY CXX_STANDARD 14)
```

Notably, if executing on Windows, it is recommended you add an additional section at the end for this file:

```
if (MSVC)
  file(GLOB TORCH_DLLS "${TORCH_INSTALL_PREFIX}/lib/*.dll")
  add_custom_command(TARGET cppapp
                     POST_BUILD
                     COMMAND ${CMAKE_COMMAND} -E copy_if_
different
                     ${TORCH_DLLS}
                     $<TARGET_FILE_DIR:cppapp>)
endif (MSVC)
```

Similarly, the following code snippet utilizes the %%writefile directive to create a .cpp file of a C++ application that loads a TorchScript model, prepares an example input, performs the prediction, and displays the predicted results:

```
%%writefile cppapp/cppapp.cpp
#include <torch/script.h>
#include <iostream>
#include <memory>

int main(int argc, const char* argv[]) {
  torch::jit::script::Module module =
torch::jit::load(argv[1]);
  std::cout << "Loaded the module\n";
  float data[] = {1, 2, 3, 4};
  torch::Tensor tensor = torch::from_blob(data, {4});
  std::vector<torch::jit::IValue> inputs;
  inputs.push_back(tensor);
  at::Tensor output = module.forward(inputs).toTensor();
  std::cout << "pred result" << output << '\n';
}
```

Next, we create a `build` folder inside the project folder (`cppapp`) with the following statement:

```
os.makedirs('cppapp/build', exist_ok=True)
```

After creating the folder, run the `cd cppapp/build` statement to set the build folder as the current working directory. We can then run the following code snippet to compile the C++ application:

```
!cmake -DCMAKE_PREFIX_PATH=/content/libtorch ..
!cmake --build . --config Release
```

We are now ready to download a traced or scripted model produced in the previous section and execute it in a C++ environment:

```
!wget <url_to/Chapter12/scripted_model.pt>
```

After downloading the model, run the following statement to execute the built application using the downloaded model:

```
!./cppapp "scripted_model.pt"
```

The previous statement should produce the following output:

```
Loaded the module
pred result 0.1100
[ CPUFloatType{1} ]
```

The TorchScript model is loaded in the C++ deployment environment successfully, and the loaded model produces the same output (0.1100) as the original PyTorch model. Similarly, we can load the traced model with the same steps. Notably, TorchLib also supports **graphics processing units** (**GPUs**); therefore, we can also perform these executions on GPUs too.

Congratulations! You have mastered the steps to export simple PyTorch models to TorchScript and execute the exported models in a C++ environment. The next section provides the code to export and deploy custom Detectron2 models using these techniques.

Deploying custom Detectron2 models

The previous section described the model formats and respective runtimes for PyTorch. It also used simple models for illustration purposes. This section will focus on deploying custom Detectron2 models into server environments using the techniques described in the previous section. This section will first describe the main export utilities that Detectron2 provides to support exporting its models. It will then provide the code to export a custom Detectron2 model into TorchScript using the tracing and scripting approaches.

Detectron2 utilities for exporting models

Detectron2 provides a wrapper class called `TracingAdapter`, which helps wrap a Detectron2 model and supports exports using the tracing method. Detectron2 models take rich inputs and also produce rich outputs (i.e., they can be dictionaries or objects). Conversely, the tracing function (`torch.jit.trace`) takes tensors as inputs/outputs. Therefore, this adapter class helps flatten the inputs of Detectron2 models as tensors so they are traceable. Additionally, it also provides the utilities to reconstruct model outputs as rich outputs from flattened results as tensors. This class takes the following input parameters:

- `model`: This is a Detectron2 model to be exported.

- `inputs`: This can be rich inputs (e.g., dictionaries or objects) and flattened by this class to produce tensors for tracing.

- `inference_func`: This is an optional callable that can be used to replace the model inference if needed.

- `allow_non_tensor`: This dictates whether inputs/outputs can contain non-tensor objects. It filters out non-tensor objects to make the model traceable but cannot reconstruct the rich inputs/outputs if enabled. The default value for this input parameter is `False`.

It has the `forward` method that wraps the original model `forward` method. This method provides flattened outputs that can be converted to rich outputs using the schema (`outputs_schema`) recorded by this class. It also provides a data member (`flattened_inputs`) to store the flattened inputs. The following is an example usage of this class:

```
# adapter can use rich inputs
adapter = TracingAdapter(model, inputs)
# trace with flattened inputs
traced = torch.jit.trace(adapter, adapter.flattened_inputs)
# produces flattened outputs
flattened_outputs = traced(*adapter.flattened_inputs)
# convert flattened outputs to rich outputs
outputs = adapter.outputs_schema(flattened_outputs)
```

The `TraceAdapter` class supports the tracing and ONNX (more on this in the next chapter) methods of exporting.

Detectron2 provides a function called `scripting_with_instances` to perform exports using the PyTorch scripting function (`torch.jit.script`). Detectron2 models use the `Instances` class, and attributes in this class can be added dynamically with eager mode. The PyTorch scripting function does not support this dynamic behavior. This is why this helper function (`scripting_with_instances`) is required. It takes the following input parameters:

- `model`: This is a Detectron2 model to be exported using the scripting method
- `fields`: This gives attribute names and types used by the `Instances` class in the model

This method creates a class called `new_Instances`, similar to `Instances` but with static attributes. The static attributes must be declared using the `fields` input parameter of the helper method. The `new_Instances` class makes the model traceable. The following is an example usage of this helper function:

```
fields = {
    "proposal_boxes": Boxes,
    "objectness_logits": Tensor,
    "pred_boxes": Boxes,
    "scores": Tensor,
    "pred_classes": Tensor,
    "pred_masks": Tensor,
    "pred_keypoints": Tensor,
    "pred_keypoint_heatmaps": Tensor,
}
model.eval()
scripted_model = scripting_with_instances(model, fields)
```

Congratulations! You now know about the two main utilities that help export the Detectron2 model with tracing and scripting methods. Let us apply these in exporting a custom Detectron2 model built previously.

Exporting a custom Detectron2 model

We first need to install Detectron2, load a dataset, and get a trained model and its configuration file for exporting:

```
# install
!python -m pip install \
'git+https://github.com/facebookresearch/detectron2.git'
# dataset
```

```
!wget -q <url_to/braintumors_coco.zip>
!unzip -q braintumors_coco.zip
# trained model
!wget <url_to/object_detector_hook.zip>
!unzip object_detector_hook.zip
output_dir = "output/object_detector_hook/"
!unzip {output_dir}model_best.pth.zip
```

The next step loads the configuration of the trained model and sets the appropriate device attribute in the configuration:

```
with open(output_dir+"cfg.pickle", "rb") as f:
  cfg = pickle.load(f)
cfg.MODEL.WEIGHTS = output_dir+"model_best.pth"
device = "cuda" if torch.cuda.is_available() else "cpu"
cfg.MODEL.DEVICE = device
```

After loading the configuration (cfg), we are ready to create a model using the DefaultPredictor class:

```
from detectron2.utils.logger import setup_logger
logger = setup_logger()
from detectron2.engine import DefaultPredictor
predictor = DefaultPredictor(cfg)
model = predictor.model
```

The next code snippet registers the test dataset for creating inputs for tracing and also for testing the exported models:

```
name_ds = "braintumors_coco"
name_ds_test = name_ds + "_test"
image_root_test = name_ds + "/test"
af = "_annotations.coco.json"
json_file_test = name_ds + "/test/" + af

from detectron2.data.datasets import register_coco_instances
register_coco_instances(
    name = name_ds_test,
    metadata = {},
    json_file = json_file_test,
```

```
        image_root = image_root_test
    )
```

The following code snippet prepares a data loader that helps to frequently load this test dataset for tracing and testing the exported models:

```
from detectron2.data import (
    build_detection_test_loader,
)
tld = build_detection_test_loader(cfg, cfg.DATASETS.TEST[0])
tli = iter(tld)
```

We will frequently load data from this test dataset, so it is sensible to create a helper method to perform this task:

```
def get_model_inputs(tli):
    inputs = next(tli)
    inputs = [{"image": input["image"] for input in inputs}]
    return inputs
```

Let us sample two batches of inputs using this helper method with the following code snippet:

```
inputs = get_model_inputs(tli)
test_inputs = get_model_inputs(tli)
```

The first batch of inputs (inputs) is used as dummy inputs to perform tracing. The second batch of inputs (test_inputs) is used to validate whether the original and exported models produce the same results given the same inputs.

Now, we are ready to create a TracingAdapter object, perform tracing, and save the traced model using the following code snippet:

```
from detectron2.export import import TracingAdapter
wrapper= TracingAdapter(model, inputs=inputs)
wrapper.eval()
traced_script_module  = torch.jit.trace(
    func                = wrapper,
    example_inputs      = wrapper.flattened_inputs)
traced_script_module.save("traced_model.pt")
```

Let us load the traced model back and perform a prediction on the `test_inputs` batch:

```
loaded_model = torch.jit.load("traced_model.pt")
loaded_model(test_inputs[0]["image"])
```

This code snippet should produce the following output:

```
(tensor([[456.4723, 163.2672, 580.1592, 317.8206],
         [456.2765, 154.8416, 584.7808, 311.3090],
         [492.0383, 161.2917, 578.7090, 254.3066],
         [484.3123, 161.7378, 575.2621, 264.6025],
         [451.0178, 200.9709, 546.3406, 313.3496],
         [461.3992, 195.3202, 546.2075, 306.3005]],
device='cuda:0',
        grad_fn=<IndexBackward0>),
 tensor([0, 1, 1, 0, 1, 0], device='cuda:0'),
 tensor([0.7608, 0.3361, 0.1744, 0.1045, 0.0825, 0.0806],
device='cuda:0',
        grad_fn=<IndexBackward0>),
 tensor([800, 800]))
```

We can validate the accuracy of this prediction by comparing it with the output from the original model:

```
predictor.model(test_inputs)
```

This statement should print the following output:

```
[{'instances': Instances(num_instances=6, image_height=800,
image_width=800, fields=[pred_boxes: Boxes(tensor([[456.4723,
163.2672, 580.1592, 317.8206],
        [456.2765, 154.8416, 584.7808, 311.3090],
        [492.0383, 161.2917, 578.7090, 254.3066],
        [484.3123, 161.7378, 575.2621, 264.6025],
        [451.0178, 200.9709, 546.3406, 313.3496],
        [461.3992, 195.3202, 546.2075, 306.3005]],
device='cuda:0',
        grad_fn=<IndexBackward0>)), scores: tensor([0.7608,
0.3361, 0.1744, 0.1045, 0.0825, 0.0806], device='cuda:0',
        grad_fn=<IndexBackward0>), pred_classes: tensor([0, 1,
1, 0, 1, 0], device='cuda:0')])}]
```

Observably, both outputs from the traced model and the original model have the same set of tensors. However, one is in rich format with instances, and the other only contains tensors. If the instances type of output is preferred, the following statement helps to perform a conversion from the tensor-only form of output to the instances type of output:

```
wrapper.outputs_schema(
    loaded_model(test_inputs[0]["image"]))
```

This statement should now produce the same output with instances as the output produced by the original model.

Similar to the tracing approach, we can prepare the fields and perform exporting using the scripting approach with the following code snippet:

```
fields = {
    "proposal_boxes": Boxes,
    "objectness_logits": Tensor,
    "pred_boxes": Boxes,
    "scores": Tensor,
    "pred_classes": Tensor,
    "pred_masks": Tensor,
    "pred_keypoints": Tensor,
    "pred_keypoint_heatmaps": Tensor,
}
model.eval()
scripted_model = scripting_with_instances(model, fields)
```

Let us also perform a prediction on the test_inputs batch using this exported model and validate the result:

```
ret = scripted_model.inference(
    test_inputs,
    do_postprocess=False)[0]
print(ret)
```

An inference result is an object of the ScriptedInstances1 class, and the predicted results are stored as attributes of this object, which we can view by executing the following statements:

```
print(ret.pred_boxes)
print(ret.pred_classes)
print(ret.scores)
```

These print statements should display the following output:

```
Boxes(tensor([[456.4723, 163.2672, 580.1592, 317.8206],
        [456.2765, 154.8416, 584.7808, 311.3090],
        [492.0383, 161.2917, 578.7090, 254.3066],
        [484.3123, 161.7378, 575.2621, 264.6025],
        [451.0178, 200.9709, 546.3406, 313.3496],
        [461.3992, 195.3202, 546.2075, 306.3005]],
device='cuda:0',
        grad_fn=<IndexBackward0>))
tensor([0, 1, 1, 0, 1, 0], device='cuda:0')
tensor([0.7608, 0.3361, 0.1744, 0.1045, 0.0825, 0.0806],
device='cuda:0',
        grad_fn=<IndexBackward0>)
```

Observably, the scripted model also produces the same output as the original model.

Congratulations! You have now mastered the utilities and steps to export custom Detectron2 models into TorchScript using tracing and scripting approaches.

> **Important note**
>
> Detectron2 also supports exporting its models to Caffe2, but Caffe2 is now deprecated, and its code base is merged with PyTorch. It also supports exporting to the TensorFlow environment, but the support has limited features. Therefore, we do not focus on these two. Additionally, ONNX is another helpful format for exporting models and deploying them to server environments. However, this format is more useful when deploying models to web and mobile environments. Thus, this approach is deferred to the next chapter.

Summary

This chapter introduced the file formats and their corresponding runtimes supported by PyTorch in general and Detectron2 specifically. It then provided simple code to export PyTorch models into TorchScript using tracing and scripting approaches. It also provided a tutorial on deploying the TorchScript model into the low-latency C++ environment at production time. This chapter then described the main utilities that Detectron2 provides to perform exports of its models to TorchScript. It then provided the steps and code to export a custom Detectron2 model into TorchScript with tracing and scripting approaches using the described utilities.

The server deployment is useful in mass and large system production. However, in many cases, there are requirements to deploy models in different environments, such as web and mobile environments. Therefore, the next chapter discusses the necessary steps to perform these deployment tasks.

13

Deploying Detectron2 Models into Browsers and Mobile Environments

This chapter introduces the **Open Neural Network Exchange (ONNX)** framework. This framework helps share deep neural networks across multiple frameworks and platforms. It is extremely helpful when there is a need to deploy Detectron2 models into browsers or mobile environments. This chapter also describes **D2Go**, a framework for training, quantizing, and deploying models with minimal memory storage and computation requirements. Models created using this framework are extremely helpful for deploying into mobile or edge devices.

By the end of this chapter, you will understand the ONNX framework and when it is helpful. You can export custom Detectron2 models into this format and deploy them into the web or browser environments. You will also understand the advantage of the D2Go framework and know how to utilize pre-trained models on its Model Zoo or train custom models using D2Go. You can also perform quantization on a custom model to further optimize its memory and improve its computation time.

Specifically, this chapter covers the following topics:

- Deploying Detectron2 models using ONNX
- Developing mobile computer vision apps with D2Go

Technical requirements

You should have completed *Chapter 1* to have an appropriate development environment for Detectron2. All the code, datasets, and results are available on the GitHub page of the book at `https://github.com/PacktPublishing/Hands-On-Computer-Vision-with-Detectron2`.

Deploying Detectron2 models using ONNX

ONNX is an open source format for representing and sharing deep learning models between different frameworks. The models can then be deployed in various platforms (e.g., servers or mobile devices) that support these frameworks. The following sections introduce ONNX and its supported frameworks and platforms, export a PyTorch model to ONNX format, and load the exported model into the browser environment.

Introduction to ONNX

ONNX aims to be a universal standard for deep learning models, allowing for interoperability between different tools, libraries, and frameworks. Microsoft and Facebook initiated the ONNX project in 2017. However, this project is currently an open source project managed by the ONNX community, which includes contributors from a wide range of organizations. This means that the project has great potential and support. The format is designed to be flexible and extensible, supporting a wide range of deep learning models, including **convolutional neural networks (CNNs)**, **recurrent neural networks (RNNs)**, and transformers.

With ONNX, deep learning models can be trained in one framework and deployed in another, making it easier to integrate deep learning models into different applications and platforms. ONNX also supports a range of optimization techniques, including **model quantization** and pruning, to help improve model performance and reduce the size of models. This feature is extremely helpful when deploying models to mobile or browser platforms.

Overall, ONNX is a powerful tool for sharing and deploying deep learning models, helping to improve the accessibility and interoperability of deep learning technology. For instance, some of the most popular frameworks that can export or import ONNX models include the PyTorch, TensorFlow, Microsoft Cognitive Toolkit (CNTK), Apache MXNet, Keras, Chainer, MATLAB, and OpenCV frameworks.

Similarly, it supports a wide range of platforms that support these frameworks. Additionally, it supports specific platforms, such as Apple Core ML, Microsoft Windows Machine Learning (WinML), NVIDIA TensorRT, and IBM Watson. Specifically, Apple Core ML is Apple's machine learning framework for iOS devices; WinML supports Microsoft's machine learning framework for Windows; NVIDIA TensorRT is NVIDIA's deep learning inference optimizer and runtime; and IBM Watson is an IBM machine learning and AI platform. The next section provides steps to export a PyTorch model to ONNX format.

Exporting a PyTorch model to ONNX

This section defines a simple PyTorch deep learning model to illustrate the steps to export a PyTorch model into ONNX format, load it back, and execute it in the ONNX environment:

```
import torch
import torch.nn as nn
```

```
class SimplePyTorchModel(nn.Module):
  def __init__(self):
    super(SimplePyTorchModel, self).__init__()
    self.linear = nn.Linear(4, 1)
    self.linear.weight.data.fill_(0.01)
    self.linear.bias.data.fill_(0.01)

  def forward(self, X):
    return torch.relu(self.linear(X))
```

Notably, we do not need to train the model but fix its weights and biases as 0.01 for demonstration purposes. The following code snippet then creates a model of this simple neural network and puts it in evaluation mode:

```
pt_model = SimplePyTorchModel()
pt_model.eval()
```

Now, we are ready to export the model into ONNX format using a dummy input:

```
dummy_X = torch.tensor([[1, 2, 3, 4]], dtype=torch.float32)
model_name = 'onnx_model.onnx'
torch.onnx.export(pt_model,
dummy_X, model_name, verbose=True)
```

To test the exported model, we first need to install the onnx package using the following statement:

```
!pip install onnx
```

After installing onnx, the following code snippet loads the exported ONNX model, checks whether it is well-formed, and prints out its graph for inspection purposes:

```
import onnx
from onnx.helper import printable_graph
# load
loaded_model = onnx.load(model_name)
# check (well formed)
onnx.checker.check_model(loaded_model)
# graph
print(printable_graph(loaded_model.graph))
```

To run the loaded model in an ONNX model, we first need to install the `onnxruntime` package using the following statement:

```
!pip install onnxruntime
```

Once the runtime environment is ready, the following code snippet helps to create an inference session and gets the input name for the loaded model:

```
import onnxruntime as ort
ort_session = ort.InferenceSession(model_name)
input_name = ort_session.get_inputs()[0].name
print(input_name)
```

The previous code snippet should display `input_name` as `onnx::Gemm_0`. This input name is important for creating an `input` object for the loaded model while performing inferencing on input data, as in the following code snippet:

```
import numpy as np
X = np.array([[2, 3, 4, 5]])
outputs = ort_session.run(
    None,
    {input_name: X.astype(np.float32)},
)
print(outputs[0])
```

The previous code snippet creates a batch input, X, with one instance of `[2, 3, 4, 5]` and performs prediction on this input. This code snippet should display the result as `[[0.15]]`. To validate the accuracy of the prediction, we can also perform prediction on the same input using the original PyTorch model as follows:

```
X = torch.tensor([[2, 3, 4, 5]], dtype=torch.float32)
with torch.no_grad():
    y = pt_model(X)
    print(y)
```

This code snippet should display the same output (`[[0.15]]`) as the loaded ONNX model executed in the ONNX inference session.

Optionally, if you are running on Google Colab, you can run the following code snippet to download this ONNX model for future use:

```
from google.colab import files
files.download('onnx_model.onnx')
```

We should now have the PyTorch model exported in ONNX format. This model can be deployed in various frameworks and platforms that support ONNX. In the next section, we will deploy this into the browser environment.

Loading an ONNX model to the browser

If the intention is to load the PyTorch or Detectron2 models into the web server environment, then the techniques described in the previous chapter should suffice. Additionally, we can also deploy ONNX models into the web server environments. However, in several cases, there might be a need to deploy deep learning models into the web browser environments. Therefore, this section provides an example of deploying the ONNX model exported in the previous section into a web browser environment. For simplicity and consistency, let us use Google Colab to execute our web application.

The following code snippet downloads the ONNX model created in the previous section to the currently running instance of Google Colab:

```
!wget <url_to/Chapter13/onnx_model.onnx>
```

Additionally, to execute the ONNX model on the browser, we need to download onnx.js to the current folder using the following snippet:

```
!wget https://cdn.jsdelivr.net/npm/onnxjs/dist/onnx.min.js
```

Once the downloads are complete, we can use the %%writefile directive to create an HTML file called load_onnx.html in the current folder. This HTML file helps to load the downloaded model and execute the model on an example input to verify the process:

```
%%writefile load_onnx.html
<html>
  <head> </head>
  <body>
    <div id="output"></div>
    <script src="onnx.min.js">
    </script>
    <script>
      const myOnnxSession = new onnx.InferenceSession();
```

```
        myOnnxSession.loadModel("./onnx_model.onnx")
        .then(() => {
            // Step 1: create an input
            // Step 2: run and update the output div
        });
    </script>
  </body>
</html>
```

This HTML file first loads the downloaded JavaScript file (onnx.js). It then creates another JavaScript to create a new ONNX inference session and loads the downloaded ONNX model (onnx_model. onnx). After the model is loaded (asynchronously), it performs the following two steps.

Step 1: This step creates an example input as an ONNX Tensor with the same data as in the previous section ([2, 3, 4, 5]):

```
const inferenceInputs = new onnx.Tensor(
    [2, 3, 4, 5],
    'float32', [1, 4]
    );
```

Step 2: This step takes the prepared input in the previous step and passes it to the inference session to perform the prediction asynchronously. Once the execution generates the output object, it then takes the output data and updates innerText of the output div:

```
myOnnxSession.run([inferenceInputs])
.then((output) => {
    const outputTensor = output.values().next().value;
    const msg = `model output tensor: ${outputTensor.data}`;
    document.getElementById('output').innerText = msg;
});
```

The following code snippet creates a helper method (expose_content) that allows exposing the content of a web server running on a given port to an iframe and embeds the generated iframe into the current Google Colab output cell:

```
from IPython.display import Javascript
def expose_content(port, height=200):
    display(Javascript("""
    (async ()=>{
```

```
        fm = document.createElement('iframe')
        fm.src = await google.colab.kernel.proxyPort(%s)
        fm.width = '95%%'
        fm.height = '%d'
        fm.frameBorder = 0
        document.body.append(fm)
    })();
    """ % (port, height)))
```

The following code snippet then starts a web server instance on port 8888 and exposes the content of the current Google Colab folder using the previously created helper method (expose_content):

```
 get_ipython().system_raw('python3 -m http.server 8888 &')
 expose_content(8888)
```

The previous code snippet should list all the files in the current working directory to the output cell of the current code cell. Clicking on the link for the previously generated HTML file (load_onnx. html) should execute it. The execution should load the model, prepare an example input, execute inference, and display the output as about 0.15 (the output may contain some numerical errors but should be approximately this correct value).

Exporting a custom Detectron2 model to ONNX

The previous chapter described the Detectron2 helper class called TracingAdapter, which helps wrap Detectron2 models so they can be traced and converted into TorchScript format. This same class is used to prepare Detectron2 models for exporting to ONNX format. This section uses this class to wrap around a custom-built Detectron2 model and exports it into ONNX format.

The code for installing Detectron2, downloading the brain tumor dataset, registering the test dataset, downloading a custom-built Detectron2 model, generating a predictor from the downloaded model, and creating a test dataset loader iterator (tli) remain the same as described in *Chapter 12* (Detectron2_Chapter12_Export.ipynb). The helper method that helps to sample input from the test dataset for tracing and validating the accuracy of the predictions made by the original Detectron2 model versus the exported model is also the same and is listed here again for ease of reference:

```
 def get_model_inputs(tli):
    inputs = next(tli)
    inputs = [{"image": input["image"] for input in inputs}]
    return inputs
```

Now, let us sample two input batches. One (`dummy_inputs`) is used as dummy inputs for exporting purposes, and another one (`test_inputs`) is used to validate whether both models (the original one and the exported one) give the same prediction results when performing inferencing on this input batch:

```
dummy_inputs = get_model_inputs(tli)
test_inputs = get_model_inputs(tli)
```

Now, we are ready to perform the export from the custom-built Detectron2 model to the ONNX format:

```
wrapper = TracingAdapter(model, inputs=dummy_inputs)
wrapper.eval()
model_name = "onnx_model.onnx"
with open(model_name, "wb") as f:
  image = dummy_inputs[0]["image"]
  torch.onnx.export(
      model   = wrapper,
      args    = (image,),
      f       = f,
      opset_version = 16
      )
```

This code snippet is straightforward. However, as of the time of writing, the current value for STABLE_ONNX_OPSET_VERSION is 11, and it does not support advanced/recent operations such as ROIAlign that this custom model uses. On the other hand, PyTorch has STABLE_ONNX_OPSET_VERSION as 16 and supports ROIAlign. Therefore, the previous code snippet changes opset_version to 16. In the future, if you see warning messages about some operations that are not supported with this ONNX opset_version, you may want to update the version and perform exports if PyTorch supports these operations.

After exporting, we can now try to load the exported model back and check whether it was exported successfully. This step requires the onnx package, and the following statement helps to install onnx on Google Colab:

```
!pip install onnx
```

You might face an encoding error if you run the installation on Google Colab (especially with GPU runtime). If so, you can run the following code to fix it:

```
import locale
def getpreferredencoding(do_setlocale = True):
    return "UTF-8"
locale.getpreferredencoding = getpreferredencoding
```

After installing the onnx package, you can run the following code snippet to load the exported model and check whether it is well-formed:

```
import onnx
onnx_model = onnx.load(model_name)
onnx.checker.check_model(onnx_model, True)
```

The next step should be performing prediction on the test_inputs batch using the loaded model. To perform prediction, we first need to install the onnxruntime package using the following statement:

```
!pip install onnxruntime
```

Now, we are ready to create an inference session and get the input name for the model:

```
import onnxruntime as ort
ort_session = ort.InferenceSession(model_name)
input_name = ort_session.get_inputs()[0].name
```

Again, the input name (input_name) is important to help prepare the input data for performing inferencing in the next step:

```
outputs = ort_session.run(
    None,
    {input_name: test_inputs[0]["image"].numpy()},
)
print(outputs)
```

This code snippet should use the exported onnx model and performs inferences on the test_inputs batch. This code snippet should print the following output:

```
[array([[456.47223, 163.26726, 580.15924, 317.82056],
        [456.27646, 154.84161, 584.78076, 311.30902],
        [492.03827, 161.29176, 578.70905, 254.3066 ],
        [484.3123 , 161.73778, 575.2621 , 264.60254],
```

```
        [451.01776, 200.9708 , 546.3406 , 313.34964],
        [461.39926, 195.32011, 546.2074 , 306.30054]],
  dtype=float32), array([0, 1, 1, 0, 1, 0], dtype=int64),
  array([0.76084214, 0.33609998, 0.1743864 , 0.1044897 ,
  0.08246017,
        0.08060681], dtype=float32), array([800, 800],
  dtype=int64)]
```

Observably, there are NumPy arrays in the output. The first array contains the predicted bounding boxes, the second includes the predicted class labels, the third has the prediction confidences (probabilities), and the last one specifies the image size. If we would like to filter the prediction with some confidence threshold, we could perform this using the third array. If you would like to have the output in Detectron2 rich format (objects of the `Instances` class), you can use the following statement to perform the conversion:

```
wrapper.outputs_schema(outputs)
```

Finally, to validate the correctness of the inference result, we can perform the same inference using the original PyTorch model as the following statement:

```
predictor.model(test_inputs)
```

The inference results should be the same as the loaded `onnx` model provides (after converting to rich outputs).

Congratulations! You now mastered the steps to export Detectron2 models into ONNX format. This model can be deployed in various platforms and frameworks that ONNX supports. Especially, the ONNX format is significantly helpful while deploying to the browser or mobile environments. However, ONNX models are still heavy and might not be feasible for deploying on mobile or edge devices. Therefore, Facebook AI Research also provides D2Go. It is a production-ready software system that supports end-to-end model training and deployment for mobile platforms. The next section discusses developing mobile computer vision apps with D2Go.

Developing mobile computer vision apps with D2Go

This section introduces D2Go, the steps to use existing D2Go models or train custom D2G models, and the model quantization.

Introduction to D2Go

Exporting Detectron2 models to TorchScript or ONNX formats may serve the purpose of deploying into the server, web, and browser environments. However, deploying Detectron2 models into mobile or edge devices may require further optimizations. Therefore, Facebook AI Research also developed D2Go, which supports training computer vision applications ready to be deployed into mobile and edge devices.

Specifically, it is a deep learning model developed on top of Detectron2 and PyTorch. Similar to Detectron2, it also contains advanced and efficient backbone neural networks that support deployments for mobile and edge devices. It provides tools and techniques for training, quantization, and deployment into mobile and edge devices. It also supports exporting to TorchScript format for deployment. Like Detectron2, it also provides a wide range of pre-trained models ready to use on its Model Zoo. Additionally, users can also train custom D2Go models on custom datasets. Let us explore these two options in the following two sections.

Using existing D2Go models

D2Go provides a set of state-of-the-art pre-trained models on its Model Zoo (https://github.com/facebookresearch/d2go/blob/main/MODEL_ZOO.md). To use these existing models, we first need to install the detectron2, mobile-vision, and d2go packages:

```
!python -m pip -qq install '<url_to/detectron2.git>'
!python -m pip -qq install '<url_to/mobile-vision.git>'
!python -m pip -qq install "<url_to/d2go.git>"
```

> **Important note**
> D2Go is under active development. Therefore, its structure and code might change from time to time, and the current code may contain bugs. This section contains some quick fixes that you may not have to do in future releases. Also, note the Python version while accessing the D2Go installation folder used in several code snippets in this chapter. You may need to change this Python version to reflect your runtime environment.

At the time of writing, the d2go package does not copy the configs folder to its installation destination. This folder is required to gather the configuration files for existing models on its Model Zoo. The following statement helps to check where the installation destination is:

```
!pip list -v | grep d2go
```

Having this output, we can check in the resulting path and see whether the configs folder exists:

```
!ls /usr/local/lib/python3.9/dist-packages/d2go
```

Note the Python version while accessing the D2Go installation folder (python3.9). If the configs folder does not exist in the installation destination, we can clone the d2go repository using the following statement:

```
!git clone <url_to/d2go.git> d2go_repo
```

After cloning this repository, we can create a folder named configs in the installation destination using the following statement:

```
d2go_dir = "/usr/local/lib/python3.9/dist-packages/d2go"
configs_dir = f"{d2go_dir}/configs"
import os
os.makedirs(configs_dir, exist_ok=True)
```

Now, we are ready to copy the contents of the configs folder from the downloaded repository over:

```
!cp -r d2go_repo/configs {d2go_dir}
```

We can list all configuration files in this folder to confirm the copy process and also check for existing models on the D2Go Model Zoo:

```
!ls "/usr/local/lib/python3.9/dist-packages/d2go/configs"
```

This statement should give the following configuration files:

```
faster_rcnn_fbnetv3a_C4_FSDP.yaml
keypoint_rcnn_fbnetv3a_dsmask_C4.yaml
faster_rcnn_fbnetv3a_C4_LSJ.yaml
mask_rcnn_fbnetv3a_C4.yaml
faster_rcnn_fbnetv3a_C4.yaml
mask_rcnn_fbnetv3a_dsmask_C4.yaml
faster_rcnn_fbnetv3a_dsmask_C4.yaml
mask_rcnn_fbnetv3g_fpn.yaml
faster_rcnn_fbnetv3g_fpn.yaml
qat_faster_rcnn_fbnetv3a_C4.yaml
```

Additionally, importing model_zoo from the d2go.model_zoo package may bring an error message like the following:

```
ModuleNotFoundError: No module named 'torch.ao.pruning'
```

If this error occurs, you can perform the following steps as a quick fix for this issue:

1. Locate the `fsdp.py` file from the d2go installation directory (e.g., on Google Colab, it is `/usr/local/lib/python3.9/dist-packages/d2go/trainer/fsdp.py`).

2. Locate and change the path to `fqn_to_module` on line 16 from `'from torch.ao.pruning import fqn_to_module'` to `'from torch.ao.sparsity.sparsifier.utils import fqn_to_module'`.

3. Press *Ctrl + S* keys to save the change.

After completing these quick fixes, we are ready to select an existing D2Go model on its Model Zoo. The ways to read and choose pre-trained models for D2Go are the same as described in *Chapter 2* for selecting pre-trained models for Detectron2. Let us select a simple model to demonstrate the steps to load an existing model on the D2Go Model Zoo:

```
from d2go.model_zoo import model_zoo
selected_model = 'faster_rcnn_fbnetv3a_C4.yaml'
model = model_zoo.get(selected_model, trained=True)
```

After this step, we should have the model loaded (`model`). Let us prepare a simple input and test the loaded model:

```
!wget -qq <url_to/800x800image/input.jpeg>
```

The following code snippet loads the downloaded image before performing inferencing on it:

```
import cv2
img = cv2.imread("./input.jpeg")
```

Like the `DefaultPredictor` class in Detectron2, D2Go also provides a class called `DemoPredictor` to generate a predictor from a loaded model. The following code snippet creates an instance of this class given the input model, performs inferencing on the loaded image (`img`), and prints the results:

```
from d2go.utils.demo_predictor import DemoPredictor
predictor = DemoPredictor(model)
outputs = predictor(img)
pred_instances = outputs["instances"]
print(pred_instances)
```

This snippet should give the following output:

```
Instances(num_instances=10, image_height=800, image_width=800,
fields=[pred_boxes: Boxes(tensor([[187.2431, 211.5772,
384.4924, 711.0326],
        [347.6918, 385.5148, 480.2947, 747.8323],
        [375.6132, 371.9870, 620.2853, 725.2900],
        [348.4673, 643.8885, 452.5689, 765.8668],
        [455.8702, 422.2232, 640.2230, 685.9368],
        [319.8739, 275.5755, 417.1956, 651.1306],
        [224.1623, 656.9628, 340.6416, 790.6293],
        [496.0885, 670.5968, 641.2301, 750.6605],
        [345.4467, 654.7966, 552.8320, 770.0833],
        [499.2968, 669.8000, 639.1102, 750.0955]])), scores:
tensor([0.9974, 0.9540, 0.9063, 0.6769, 0.4782, 0.3086, 0.2689,
0.1755, 0.0990,
        0.0933]), pred_classes: tensor([ 0,   0,   0, 30, 33,   0,
30, 30, 30, 31])])
```

Observably, all detected objects are listed, regardless of predictive confidence. Let us define a threshold (`thres_test = 0.5`) and filter the predicted results using this value:

```
thresh_test = 0.5
selected_idxs = pred_instances.scores > thresh_test
pred_instances = pred_instances[selected_idxs]
pred_instances
```

This snippet filters for instances with high predictive confidence:

```
Instances(num_instances=4, image_height=800, image_width=800,
fields=[pred_boxes: Boxes(tensor([[187.2431, 211.5772,
384.4924, 711.0326],
        [347.6918, 385.5148, 480.2947, 747.8323],
        [375.6132, 371.9870, 620.2853, 725.2900],
        [348.4673, 643.8885, 452.5689, 765.8668]])), scores:
tensor([0.9974, 0.9540, 0.9063, 0.6769]), pred_classes:
tensor([ 0,   0,   0, 30])])
```

Let us visualize the outputs and quantitatively evaluate the prediction result:

```
from detectron2.utils.visualizer import Visualizer
```

```
from detectron2.data import MetadataCatalog
metadata = MetadataCatalog.get("coco_2017_train")
v = Visualizer(img[:, :, ::-1], metadata)
instances = pred_instances.to("cpu")
annotated_img = v.draw_instance_predictions(instances)
imshow(annotated_img.get_image()[:,:,::-1])
```

Figure 13.1 shows the visualizations of the predicted results overlaid on top of the input image:

Figure 13.1: Sample prediction results from an existing D2Go model

Observably, the existing model performs pretty well. However, there is an obvious accuracy trade-off for the system's simplicity and light weight (one person is not detected).

These D2Go existing models are state-of-the-art models pre-trained on popular datasets. These models should meet the most common need for computer vision applications on mobile and edge devices. However, if there is a custom dataset, D2Go also allows training its models on custom datasets for specific business requirements.

Training custom D2Go models

Let us take the brain tumor dataset and train a D2Go model to perform brain tumor detection and classification tasks. Similar to the previous section, we need to install the detectron2, mobile-vision, and d2go packages. We must also copy the configs folder to the d2go installation destination and change the line for importing fqn_to_module. These steps remain the same as described in the previous section.

We also need to download the brain tumor dataset and register the train and test datasets to Detectron2 as we have often done while training Detectron2 models (the code remains the same). After downloading the datasets and registering them, let us select a D2Go model for training. For this task, D2Go provides a class called GeneralizedRCNNRunner to help create a configuration file and train custom models. Let us start by importing this class, selecting a model, and creating an instance of this runner class:

```
from d2go.model_zoo import model_zoo
from d2go.runner import GeneralizedRCNNRunner
selected_model = "faster_rcnn_fbnetv3a_C4.yaml"
nc = 2
output_dir = './output'
model_url = selected_model
runner = GeneralizedRCNNRunner()
```

Now, we are ready to get a configuration for this model before starting training:

```
cfg = runner.get_default_cfg()
cfg.merge_from_file(model_zoo.get_config_file(model_url))
cfg.MODEL_EMA.ENABLED = False
cfg.DATASETS.TRAIN = (name_ds_train,)
cfg.DATASETS.TEST = (name_ds_test,)
cfg.DATALOADER.NUM_WORKERS = 2
cfg.MODEL.WEIGHTS = model_zoo.get_checkpoint_url(model_url)
cfg.MODEL.DEVICE ="cpu" if ('CI' in os.environ) else "cuda"
cfg.SOLVER.IMS_PER_BATCH = 2
cfg.SOLVER.BASE_LR = 0.00025
cfg.SOLVER.MAX_ITER = 5 if ('CI' in os.environ) else 5000
cfg.SOLVER.STEPS = []
cfg.MODEL.ROI_HEADS.BATCH_SIZE_PER_IMAGE = 128
cfg.MODEL.ROI_HEADS.NUM_CLASSES = nc
cfg.OUTPUT_DIR = output_dir
os.makedirs(cfg.OUTPUT_DIR, exist_ok=True)
```

Most of the configuration parameters are the same as those for the Detectron2 model: creating a configuration, merging it with an existing configuration file, and downloading pre-trained weights. We are now ready to build a model using the configuration (cfg) and train this model using the brain tumor dataset:

```
model = runner.build_model(cfg)
runner.do_train(cfg, model, resume=False)
```

Observably, this model should take a lot less time to train compared to training the Detectron2 model because it is a lightweight model. Let us evaluate the trained model on the test dataset:

```
metrics = runner.do_test(cfg, model)
print(metrics)
```

The output could be different from run to run due to randomness during the training process. However, the **Average Precision (AP)** at **Intersection over Union (IoU)** should be about 0.48, which is a reasonably good evaluation result. We can use code similar to the previous section to make predictions on some sample data and visualize the results to evaluate the performance of the trained model qualitatively.

This trained model is lightweight and meant to be deployed on mobile environments. However, it still uses 32-bit floating point numbers to store its weights and activations; therefore, it is still considered inefficient. If the model space needs to be compressed even more, the model quantization technique comes in handy. The next section discusses this technique in detail.

Model quantization

Model quantization is a technique that helps to reduce computational complexity and memory footprint with a little trade-off of accuracy for deep neural networks. Specifically, deep learning models often store their weights and activations in 32-bit floating-point numbers, and this high-precision number format requires a large amount of data storage and computation time. Therefore, model quantization tackles this problem by converting the weights and activations to lower precision data types, such as 8-bit integers. This reduction helps reduce memory usage and computation time. Notably, while quantization can significantly reduce the memory and computation requirements of a model, it can also cause a loss in model accuracy.

Continuing from the previous section, let us start by printing out the data type for the first parameter of the trained model using the following code snippet:

```
for name, param in model.named_parameters():
    if param.requires_grad:
        print(name, param.data.dtype)
        break
```

This snippet should display the following output:

```
backbone.body.trunk0.fbnetv2_0_0.conv.weight torch.float32
```

Observably, its parameters are stored in a `float32` data type, and performing quantization should help reduce the memory and computation loads:

```
# some import statements are removed for space efficiency
pytorch_model = model.to("cpu")
pytorch_model.eval()
test_loader = runner.build_detection_test_loader(
            cfg, dataset_name=name_ds_test
        )
predictor_path = convert_and_export_predictor(
        cfg             = cfg,
        pytorch_model   = copy.deepcopy(pytorch_model),
        predictor_type  = "torchscript_int8",
        output_dir      = './',
        data_loader     = test_loader
    )
```

Specifically, this code snippet converts the D2Go model to CPU and sets it in evaluation mode. It then converts this model into the `torchscript_int8` format.

Let us load the quantized model and perform an evaluation on the test dataset and compare the results with the evaluation results produced by the original D2Go model:

```
from mobile_cv.predictor.api import create_predictor
int8_model = create_predictor(predictor_path)
metrics = runner.do_test(cfg, int8_model)
print(metrics)
```

The results may change from run to run due to randomness during training. However, this model should generally have a smaller memory requirement and faster inference time with a little trade-off in accuracy.

Congratulations! You have mastered the steps to use existing D2Go models and train D2Go models on custom datasets. Additionally, you should understand the benefits that the quantization technique brings and the trade-off for those benefits.

Summary

This chapter introduced the ONNX model format and the platforms and frameworks that support it. This framework helps Detectron2 models to be interoperable with different frameworks and platforms. It then provides the steps to export Detectron2 models to this format and the code to deploy the exported model in the browser environments. This chapter also introduced D2Go, a framework for training, optimizing, and deploying neural networks for computer vision applications with minimal memory storage and computation resources. Additionally, its models are prepared to be further optimized using the quantization technique, which converts the model weights and activations in lower-precision number systems. This quantization step further reduces the model memory requirement and improves computation performance. Therefore, D2Go models are suitable for deploying into mobile or edge devices. D2Go also has pre-trained models on its Model Zoo. Thus, this chapter provides the steps to build computer vision applications using existing models on the D2Go Model Zoo. It also provides the steps to train custom D2Go models and perform quantization using D2Go utilities.

Congratulations, and thank you for following along until the last chapter! You should now consider yourself an expert in developing practical computer vision applications using Detectron2. Additionally, the theory provided in this book should help you to explore other computer vision deep learning architectures such as **You Only Look Once (YOLO)**.

Index

Packtpub.com

Subscribe to our online digital library for full access to over 7,000 books and videos, as well as industry leading tools to help you plan your personal development and advance your career. For more information, please visit our website.

Why subscribe?

- Spend less time learning and more time coding with practical eBooks and Videos from over 4,000 industry professionals

- Improve your learning with Skill Plans built especially for you

- Get a free eBook or video every month

- Fully searchable for easy access to vital information

- Copy and paste, print, and bookmark content

Did you know that Packt offers eBook versions of every book published, with PDF and ePub files available? You can upgrade to the eBook version at packtpub.com and as a print book customer, you are entitled to a discount on the eBook copy. Get in touch with us at customercare@packtpub.com for more details.

At www.packtpub.com, you can also read a collection of free technical articles, sign up for a range of free newsletters, and receive exclusive discounts and offers on Packt books and eBooks.

Other Books You May Enjoy

If you enjoyed this book, you may be interested in these other books by Packt:

Modern Time Series Forecasting with Python

Manu Joseph

ISBN: 9781803246802

- Find out how to manipulate and visualize time series data like a pro
- Set strong baselines with popular models such as ARIMA
- Discover how time series forecasting can be cast as regression
- Engineer features for machine learning models for forecasting
- Explore the exciting world of ensembling and stacking models
- Get to grips with the global forecasting paradigm
- Understand and apply state-of-the-art DL models such as N-BEATS and Autoformer
- Explore multi-step forecasting and cross-validation strategies

Deep Learning and XAI Techniques for Anomaly Detection

Cher Simon

ISBN: 9781804617755

- Explore deep learning frameworks for anomaly detection
- Mitigate bias to ensure unbiased and ethical analysis
- Increase your privacy and regulatory compliance awareness
- Build deep learning anomaly detectors in several domains
- Compare intrinsic and post hoc explainability methods
- Examine backpropagation and perturbation methods
- Conduct model-agnostic and model-specific explainability techniques
- Evaluate the explainability of your deep learning models

Packt is searching for authors like you

If you're interested in becoming an author for Packt, please visit `authors.packtpub.com` and apply today. We have worked with thousands of developers and tech professionals, just like you, to help them share their insight with the global tech community. You can make a general application, apply for a specific hot topic that we are recruiting an author for, or submit your own idea.

Share Your Thoughts

Now you've finished *Hands-On Computer Vision with Detectron2*, we'd love to hear your thoughts! Scan the QR code below to go straight to the Amazon review page for this book and share your feedback or leave a review on the site that you purchased it from.

`https://packt.link/r/1-800-56162-8`

Your review is important to us and the tech community and will help us make sure we're delivering excellent quality content.

Download a free PDF copy of this book

Thanks for purchasing this book!

Do you like to read on the go but are unable to carry your print books everywhere?

Is your eBook purchase not compatible with the device of your choice?

Don't worry, now with every Packt book you get a DRM-free PDF version of that book at no cost.

Read anywhere, any place, on any device. Search, copy, and paste code from your favorite technical books directly into your application.

The perks don't stop there, you can get exclusive access to discounts, newsletters, and great free content in your inbox daily

Follow these simple steps to get the benefits:

1. Scan the QR code or visit the link below

https://packt.link/free-ebook/9781800561625

2. Submit your proof of purchase
3. That's it! We'll send your free PDF and other benefits to your email directly